THE
STARGAZER'S
GUIDE

ABOUT THE AUTHOR

EMILY WINTERBURN comes from a background in physics and the history of science. As the Curator of Astronomy at the Royal Academy Observatory, Greenwich, she was responsible for one of the world's most important astronomy collections. Emily has written for the BBC and *Astonomy Now* magazine and has appeared on the BBC's "What the Ancients Did for Us," the Channel 4 News, and "In Our Time with Melvyn Bragg." She lives in Yorkshire, England, with her husband and two children and works at the Leeds University History of Science.

THE STARGAZER'S GUIDE

How to Read Our Night Sky

EMILY WINTERBURN

HARPER PERENNIAL

NEW YORK • LONDON • TORONTO • SYDNEY • NEW DELHI • AUCKLAND

HARPER PERENNIAL

First published in the United Kingdom by Constable & Robinson Ltd, 2008.

THE STARGAZER'S GUIDE. Copyright © 2008 by Emily Winterburn.
Starcharts © 2008 by Greg Smye-Rumsby, www.conceptdesign.co.uk.
All rights reserved. Printed in the United States of America. No part of this
book may be used or reproduced in any manner whatsoever without written
permission except in the case of brief quotations embodied in critical articles
and reviews. For information address HarperCollins Publishers,
10 East 53rd Street, New York, NY 10022.

HarperCollins books may be purchased for educational, business, or sales
promotional use. For information please write: Special Markets Department,
HarperCollins Publishers, 10 East 25th Street, New York, NY 10022.

FIRST HARPER PERENNIAL EDITION PUBLISHED IN 2009.

Library of Congress Cataloging-in-Publication Data
is available upon request.

ISBN 978-0-06-178969-4 (pbk.)

09 10 11 12 13 ❖/RRD 10 09 8 7 6 5 4 3 2 1

For Lottie and Sam

Contents

Acknowledgements

Although my name has to go on the cover, I cannot claim sole credit for the content of this book. For the images I am very grateful to a number of stargazing photographers who were kind enough to allow me to use their pictures: I would especially like to thank Michael Oates, Robert Gendler and Ray and Barnaby Norris. I would also like to give a special thanks to Greg Smye-Rumsby for agreeing to illustrate the book and in particular for his fantastic star charts. For helping to shape the original idea and then improve the text my thanks go to Diane Banks my agent, Becky Hardie my editor at Constable, Angela Ayton, Robert Massey (at the Royal Astronomical Society) and Elizabeth Stone. Finally I would like to thank my parents, Angela Ayton and Michael Winterburn and Bob and all the staff at our local nursery for looking after my children and so making it possible for me to write.

Starry, Starry Night

A STARRY SKY, LIKE THE OCEAN, has the capacity to fill us with wonder. When we stand in a field, the mountains, a desert or even by the sea and look at the stars without the interference of street lighting it is easy to feel overwhelmed. The whole vast universe is in front of us – or at least as much of it as the naked eye can see. This is in itself both beautiful and fascinating. But a little knowledge of the history and science of the night sky can help us appreciate it even more.

Stargazing is in many ways an exercise in visual history. When we look at the stars, we are actually looking at the past. The patterns we use today to navigate our way around the sky come not from modern research but from stories created by earlier cultures – our understanding of individual stars comes from centuries of accumulated research and evolving stories. More than this, we only see the stars as they looked when the light now reaching our eyes left them: we cannot see them as they are at the moment. Aldebaran in

Taurus, for example, a star sixty-five light years away, looks to our eyes as it appeared sixty-five years ago.

There have always been stargazers, people interested simply in knowing a little more about what they can see in the night sky. Indeed, it seems to be an almost universal human pastime to look up and wonder at the stars. Astronomy, the more scientific branch of this activity, was for a long time viewed as the practical arm of astrology. While astronomers kept careful and accurate records of the position of every star, planet and comet in the sky, astrologers interpreted those data to predict the future for leaders and later for the paying public. As astronomers in the late seventeenth and eighteenth centuries began to discredit astrologers' claims, astronomy lectures, books and games became popular as a form of 'rational recreation'. It became fashionable to spend one's evenings learning the names and stories associated with the stars and constellations (or star patterns) and to discuss the latest astronomer's new theory or discovery. Astronomers were celebrities and attending lectures was the height of fashion.

I love the idea that having some knowledge of science, and astronomy in particular, was once so hip. As a rather unfashionable adolescent I spent many evenings sitting through science lectures at Birkbeck College and the Royal Institution, very much a fringe activity at the time. It was only much later I discovered that in the 1790s I might have been considered right on trend.

Astronomy, as one of the oldest sciences, has a long and appealing history. So for me, this book is an opportunity to delve into past ways of telling the many stories associated with the sky, to tell them for a twenty-first-century audience, bringing the joys of stargazing to a whole new generation.

At university I studied physics, with a little astronomy thrown

This print, entitled 'The Kentish hop merchant and the lecturer on optics', dates from the early 1800s.

in. As curator of an astronomy collection (at the Royal Observatory in Greenwich) I have spent the last ten years learning about and explaining to others what there is to see in the sky, why it's interesting and how previous generations have viewed and interpreted it.

Just recently, I've been working on an exhibition about telescopes and their enduring attraction. As part of the research I was discussing the project with an artist who has worked on a number of astronomical installations. He pointed out that, in many ways, people generally know less now about astronomy than ever before. We live in cities and so rarely see many of the stars, much less the

Milky Way. Our everyday experience of power sources comes not from tides and the flow of rivers (which are in turn tied to the movements of the Moon) but from batteries and mains electricity; and we perhaps no longer seriously believe the stars and planets have any influence over our physical or mental well-being. But, both the artist and I agreed, things do seem to be changing. For a variety of reasons, we're all more interested in our environment and many of us are taking real steps to re-engage with our natural world. Some knowledge of the stars, of their relationship to the seasons, to time and the natural rhythms of day and night, is an integral part of that undertaking. We may soon see the return of astronomy – or at least stargazing – as a popular pursuit.

Today we tend to think of the night sky as something beautiful to look at, something to be explored and, if the number of newspaper horoscopes is anything to go by, somewhere that reveals the future. Historically, the night sky has had more specific uses. It has been used to create and then regulate calendars and timepieces, to navigate on land and at sea and to aid medical diagnosis and treatment.

Ancient cultures used what they saw to tell stories that would explain how the Earth, sky and human life came into existence. Later cultures used the night sky to create stories about how the gods taught us to behave and how they shaped the details of the world we live in. Today modern astronomers use various tools to look into space and deep into the individual stars and star groupings before giving us scientific explanations – what stars consist of, how they were made, how they will end, how they move – all of which informs the way we look at ourselves and the world.

Alongside the tales of creation and the interaction of gods and mortals came systematic record keeping. It was noted how the rising and setting of different stars and planets (or 'wanderers') related to the agricultural year. Gradually links were made; so, for example, the ancient Egyptians knew that with the pre-dawn rising of the star Sirius in Canis Major came the annual flooding of the Nile. Later the regular movements of the Sun and the stars were used to divide up the day and night into smaller units, then to tell the time, using sundials in the day and nocturnals and astrolabes at night.

As ships ventured into open oceans away from landmarks, the stars became increasingly important as the sailors' only means of navigation. Knowledge of the stars gave them their basic north, south, east and west co-ordinates. The stars could also, when used with the right tables, give them their longitude – the distance east or west of an agreed point or meridian, today universally agreed to be Greenwich in London. On land, too, the stars have been used as a way of navigating deserts or other wildernesses without landmarks. Early scientific instruments show that the stars were also used, in the absence of a compass, to help travelling Muslims find Mecca.

Both timekeeping and navigation gave astronomy its purpose and justified much of its early state funding, but it was astrology that gave it its greatest boost. From the ancient Greeks onwards, medicine and astrology had been closely linked: the planets were thought to alter the balance of the humours and the zodiac signs were said to govern different parts of the human body, from Aries at the head to Pisces at the feet. Diagnosis was based on casting a horoscope, while treatment was based on selecting plants and minerals with the right astrological associations.

By the first century CE it was not just the physical body that was thought to be ruled by the stars and planets but, as a natural extension of this theory, the mind and personality, or temperament. Theories developed that related key episodes in world history with astrologically significant changes in the stars. Thus, the history of the human race itself became part of the same relationship between man and the heavens. Then there were personal horoscopes, which kept many respected astronomers as well as unregulated charlatans in work for many years. Today's astronomers do not cast horoscopes, but four hundred years ago it was for many a means of keeping solvent. Johann Kepler, now famous for several important astronomical and mathematical theories including Kepler's Law, is known to have cast around four hundred horoscopes in his time.

The aim of this book is to bring the sky to life. I have therefore grouped the constellations by month rather than, as has become common practice in amateur astronomy books, in alphabetical order. This book is not strictly for the amateur astronomer – although they are welcome to read it. It is not a conventional, text-book guide that will allow the constellations to be methodically ticked off as they are spotted. Instead, you can use the charts to find Ursa Major (otherwise known as the Great Bear, the Plough or the Saucepan) and then you might be interested in finding that bear's son, Ursa Minor. You might also wish to follow the herdsman, Boötes, as he guides the bears in their journey around the pole star with his hunting dogs, Canes Venatici. You might then want to know where these constellations come from and the story that links them together.

The Greek myths associated with the constellations are only the beginning of what you'll find in this book. In total there are eighty-eight internationally recognized constellations in the night sky. Only forty-eight of these were described by the ancient Greeks; what we've learned since about the rest is fascinating. Then there is our Sun, our closest star. Despite the extreme hazards associated with looking directly at the Sun with the naked eye, our local star offers stargazers, amateur and professional astronomers alike, the opportunity of observing a star close up. Eclipses, in particular, present a way of examining the outer layers of this type of star as well as the perfect excuse to travel to obscure corners of the globe in order to get the best view.

The Moon and planets I mention only in passing: this is, after all, a stargazer's guide, and these are not, nor have they ever been, described as or confused with stars (as comets and meteors have). The other reason they are not included is because they do not appear and disappear at regular yearly intervals and so do not lend themselves easily to a month-by-month guide.

The Moon moves in an erratic path across the sky, coming back to the same place only once every eighteen years. Harder to see are the planets, which at first glance look remarkably like the stars. They, like the Earth, move around the Sun, but each does so at its own speed. This means that they do not keep in step with the constellations but rather appear to move through them: they do not actually move between the stars but from Earth appear to weave in and out of the star groupings. The stars are much further away. But this guide does include comets, meteor showers (or shooting stars) and satellites, all with their own stories – how they were discovered and interpreted by previous generations and by other cultures or, in the case of artificial satellites, why they are there at all.

Astronomy has always been one of the more popular sciences, since the stars retain a certain romantic charm. Nursery rhymes have been written about them and poems dedicated to them, while cinema has often used stargazing as shorthand for introducing a particularly imaginative, idealistic and attractive character. Today the card games, board games and toys so popular in the eighteenth and nineteenth centuries are now mostly found in museums. The few astronomy games and books, lectures and societies that do appear tend to be aimed at a much narrower, more dedicated 'amateur astronomy' audience. It is an attractive hobby, as astronomy is one of the few areas of modern science where amateurs can and do still make important discoveries. As recently as 1995 an American amateur astronomer, Thomas Bopp, with the aid of his friend's telescope, discovered a comet simultaneously with professional astronomer Alan Hale, giving us the comet Hale-Bopp. However, this kind of highly dedicated amateur astronomy offers little scope for those with a layman's interest in the romance of the sky.

This book is written for such a layman, a stargazer. A stargazer is anyone who looks at the stars and would like to know more about what they are seeing. Stargazing requires no equipment, except perhaps something comfortable to sit on and a star map (such as the ones in this book). For most of us, it is a leisurely activity undertaken occasionally, generally when we find ourselves with time on our hands, in some out-of-the-ordinary place. City lights and cloudy weather mean it is mostly reserved for holidays in the country or at the seaside, or even the occasional adventure into the desert.

Away from city lights there are so many stars to see that the rationale behind organizing them into constellations becomes immediately apparent. This is how you find your way around the sky.

However, from a city sky it is possible to make out a small number of bright stars which make up constellations. Ursa Major and Orion, for example, are both made up of a number of very bright stars and can often be seen even from the most built-up areas.

Wherever you are in the world, stars will appear to move with all the other stars (though, of course, in reality it is we who are moving). Planets seem to wander through the constellations, at apparently different speeds depending on how close they are to the Sun and how fast they are moving. Comets will appear and disappear over the course of a few months. Satellites will appear and disappear in minutes, while meteors or shooting stars seem to streak across the sky in seconds. This book will help you tell the difference between these phenomena, explain how to interpret what you see and where best to view them.

In the following pages we travel through the sky month by month, meeting the constellations visible that month and the stars within them. Each has a story, whether it is a Greek myth, a historical tale of discovery or an astrological prediction. By the end, you should feel well prepared to make the most of any cloud-free night sky. It may even inspire in you the same passion for the sky as all those generations of stargazers before you.

CHAPTER ONE

April and the Bears

I'T'S DIFFICULT TO SAY WHEN I became a stargazer. I don't remember learning the constellations Ursa Major, Ursa Minor (the two bears) or Orion, so I suppose I must have been quite young. I do remember trying to spot Halley's Comet when I was about eleven on its much-talked-about return in 1986, but I don't remember that interest turning into a full-blown passion for astronomy. But then stargazing is not exactly a hobby; it's more an ongoing curiosity. It's like having a living museum that travels around with you, its exhibitions revealing valuable secrets about us as people and as part of a world culture.

Stargazing can also be a quieter, more domestic activity. It's probably something you already do. In winter, as the nights draw in, you may even find yourself stargazing on your way home from work. You can probably already identify some of the sky's more well-known constellations. Take the northern hemisphere's Great Bear, Ursa Major, otherwise known as the Plough, the Saucepan or the Big Dipper. Or take the southern hemisphere's Crux or South-ern Cross. Both are ideally suited for all kinds of observing, whether from the remotest corner of the Earth or at the heart of one of its busiest cities. They are a good place to start. Their stories draw in less familiar constellations, and details of the myths that built up around them give us clues as to how the heavens appeared to behave, which in turn leads us into modern explanations for the same phenomena. The range of ideas, problems, histories and scientific concepts they introduce are a good basis from which to start our year's worth of stargazing.

We begin in April, when these most familiar constellations are usefully central in the sky, as a nod to the ancients' and astronomy's debt to astrology. Astrologers began their year in spring (and still do – look at most horoscope pages and you will see they begin with Aries). The spring has, for very good reasons, traditionally been associated with new beginnings, with new life and the start of the agricultural year. One reason early civilizations studied the stars and grouped them into constellations was to enable them to predict the start of each new season, beginning with the spring. The zodiac constellations were key to this, and many of our depictions of each sign still have strong associations with their season: Virgo holding an ear of corn to symbolize the fruits of summer; Aries symbolized by the ram, a traditional symbol of male fertility and so the begin-nings of new life. Theoretically, a year can begin anywhere – today's Gregorian calendar means our year begins on 1 January but this is an arbitrary choice. But since this book is about stargazing and its heritage it seems fitting to take April, just after the north-ern hemisphere's spring equinox, as the start of our year.

Dark skies and city observing

I have always lived in cities. City observing can be problematic for the stargazer because the light produced by street lights and a high density of buildings, all with their own lights, washes out all but the brightest stars. Even on the clearest nights, you see only a tiny fraction of the stars that should be visible.

Both astronomers and environmental groups run various campaigns that tackle the problem of light pollution, but they still have a long way to go. There are ways of controlling the problem of bright cities though, as any dark skies campaigner will tell you. In Tenerife and La Palma in the Canary Islands, for example, which together house the various telescopes making up the European Northern Observatory, laws are in place to protect the astronomers' dark skies. There, street lighting is designed to point downwards so that it lights only the street, not the skies. There are also laws regulating other types of outdoor lighting such as billboards. But dark skies are not just for astronomers. Environmental campaigners similarly argue for dark skies on the grounds of energy consumption. Light nights also have an effect on the body clocks of birds, animals and insects, making them vulnerable to predators and so shifting the whole balance of the local ecological system.

NASA's satellite image of the Earth at night gives you an idea of the extent of the problem. (In fact, this image comprises several images brought together in order to show how each country looks at night compared with another.)

Broadly speaking, you can see from this that North America, Europe and Japan are terrible places to observe from. But, more than that, you can see just how bright places with lots of cities

This composite image of the whole world at night illustrates just how great the problem of light pollution is in our cities.

are. This becomes even clearer if you look at a map of a specific country.

In the Campaign to Protect Rural England (CPRE) map, below, of the UK you can see the areas around London, Leeds, Manchester and Liverpool – the big cities – flooded with light. The more remote areas, particularly parts of northern Scotland and Wales, have almost no light pollution and as a result are fantastic places for unhampered stargazing.

This means that for some observing you need to think about where might be the best location. In cities you can see the brightest stars, planets, sometimes comets and satellites, but not much else.

This image of the UK shows how bright our cities are at night.

In the remotest desert or in the middle of the ocean you will see an unbelievable number of stars on a clear night, though it may then be difficult to make out familiar patterns. Then, perhaps most practically, you have the countryside and the seaside, away from big cities. These include campsites and festivals, which are often situated away from built-up highly lit areas. It is in these kinds of places that you can make your best attempt at seeing some of the more obscure constellations.

Ursa Major

Despite all this, our two main constellations for April, Ursa Major and Crux, are visible even in cities. Ursa Major is one of the oldest constellations and its ancient story is apparent in the name we still use to describe it − the Great Bear. In the April sky Ursa Major should be directly overhead. Take a look at the star chart for spring and you will find Ursa Major towards the centre. Once you've found this familiar shape, follow the two stars that make up the back of the bear (from Merak to Dubhe) up to a bright star. This is Polaris. For the time being, this is the northern pole star around which the whole sky appears to rotate. Because of something called precession, this will not always be the case. Precession is, roughly speaking, the wobble of the Earth (think of a spinning top), which means that, over long periods of time, the star directly above the North Pole will change. The wobble is caused by the gravitational pull of the celestial bodies that surround the Earth, most notably the Sun and Moon.

In addition to being our current pole star, Polaris is also the tip of the tail of Ursa Minor (the Little Bear). To find Boötes (the herdsman), we need now to return to Ursa Major. By following the tail as it curves round you will come to another bright star. Arcturus is the brightest star in Boötes and above it, with the star chart to help you, you can make out the rest of the constellation. Finally, Boötes's hunting dogs, Canes Venatici, can be found just under the tail of Ursa Major.

Bright stars

Boötes's Arcturus is one of the brightest stars in the night's sky. The name Arcturus comes from the Greek Arktourus, meaning 'Bear Guard'. It has an apparent magnitude of −0.05, making it brighter than all the stars except Sirius (mag. −1.47) in the constellation Canis Major, and Canopus (mag. −0.72) in Carina. The brightness of stars, as they appear from Earth, is measured in apparent magnitude in a scale going from minus numbers for the brightest stars to very high numbers for stars so faint they can only be seen with the most powerful telescopes. The scale uses the star Vega in the constellation Lyra as its baseline (we'll come back to this star and this constellation in the next chapter). Vega has an apparent magnitude of 0; anything brighter than this has an apparent magnitude measured in negative numbers − the Sun, for example, has an apparent magnitude of −27. Anything dimmer has a positive number, apparent magnitude increasing as the star gets dimmer. With good eyesight the naked eye can just see up to about magnitude 6, though stars have been measured with high-powered telescopes as having apparent magnitudes of up to 38.

Delving into the details of some of our brightest stars can be a straightforward way of understanding the different types of stars. Stars are described in a number of ways, one of the most common of which is to categorize them according to where they are on their lifecycle. Stars begin (if you can begin a circle) as nebulae. These are clouds of dust and gas which condense down to become stars of different sizes. To begin with they are protostars, balls of dust and gas, held together and made to spin by gravity. They heat up and eventually become hot enough to start at their core a process called nuclear fusion, whereby atoms of one type of element (or more accurately atomic nuclei since the atoms in a star's core will have been ionized) are made into atoms of another type, releasing energy in the form of electromagnetic radiation (which includes heat and light). Since these protostars are made up mostly of hydrogen, it is this that is fused into another type of atom, helium, in the star's core. Once this process has begun, and radiation pressure pushing out balances gravity's inward pull, the protostar is classed as a fully formed 'main sequence' star. Our Sun is a main sequence star, as is Vega. Eventually most (though not all) of the hydrogen in its core is used up and the star becomes a red giant (or supergiant for stars very much more massive than our Sun) as its outer layers expand outwards and as it starts turning the helium into carbon at its core.

Another reason for describing these types of stars as main sequence is because of their position on the Hertzsprung–Russell diagram (see page 157). (This graph, produced in 1910 by astronomers Ejnar Hertzsprung from Denmark and Henry Norris Russell from America, relates the brightness of a star to its temperature. From this and its mass you can tell a star's past and future.) These types of stars, when plotted on this graph, fall within the main curve or sequence of it. We will come back to this graph in later chapters because it can help us to better visualize the relationship between stars at different stages of their lifecycle or evolution. For now, though, it's enough just to familiarize ourselves with some of the terms.

Stars that begin as very massive stars (anything from around six to 150 times the mass of our Sun) have relatively short lives, ending up as supernovas once they've finished creating new elements at their core by nuclear fusion. As supernovas they throw off their outer layer in a dramatic explosion, leaving their core to become a neutron star or even a black hole. Stars that begin as low mass stars last much longer, living for billions rather than millions of years. Low or mid-mass stars, like our Sun, similarly throw off their outer layer. What is left is a core surrounded by a glowing shell, a planetary nebula. Eventually the core becomes a white dwarf and then finally, as it cools, a black dwarf. In both cases their outer layer contains a large amount of hydrogen – this becomes a nebula bringing us back round to the beginning and the birthplace of stars. Very low mass stars, those typically less than half the mass of our Sun, become red dwarfs where they undergo very slow nuclear fusion in their cores, and so live extremely long lives.

Arcturus is a red giant, which means it is a low mass (but not very low mass) star, coming towards the end of its life. It has finished converting hydrogen to helium at its core and its outer layers have expanded and cooled. It is expected that next the core will contract and heat up and the helium inside it will start to be converted into carbon and oxygen; this releases energy in the form of electromagnetic radiation (which includes heat and light). The shell of gas around the core, which will begin to glow, is known as a planetary nebula since it is spherical and so shaped as a planet. The name is based purely on the shape and should not be confused with anything to do with planet formation. Once this nebula drifts away, all that is left is the core, no longer converting helium into heavier elements, in fact no longer producing any energy at all. Initially it is still very hot and glows white. At this stage it is a white dwarf.

As it cools, its brightness fades. Our Sun, when it uses up all the hydrogen in its core, in some 5 billion years, is expected to follow more or less the same path.

At only 36.7 light years away, Arcturus is relatively close to the Earth. This, coupled with the amount of light it actually produces, helps to account for its brightness. Ursa Minor's Polaris, on the other hand, is around 400 light years away. Yet despite this it is often thought to be the brightest star in the northern hemisphere, if not the sky. In fact it is only the forty-eighth brightest, with an apparent magnitude of 1.97. Perhaps it is considered brighter because it is always high in the sky for most observers in the northern hemisphere and consequently always quite easy to spot. Polaris is a yellow supergiant. Its colour is an indication of its temperature − red stars are relatively cool, at around 3,000K; blue very hot, around 50,000K; yellow somewhere between the two. (K stands for Kelvin, the standard scientific measure of temperature. It's basically the same as the more familiar centigrade or Celsius scale, except that where 0°C marks the freezing temperature of water, 0K (equal to −273°C) marks the freezing degree of everything. Nothing in the universe is colder than 0K.) In some instances the colour of a star is just about detectable with the naked eye but often it's something we have to take on trust. Polaris is a very large star, with a mass of around six of our Suns, and so consequently became a supergiant rather than a giant when it began turning the helium at its core into carbon.

The constellations these bright stars belong to form part of a group that together represent the characters, and to some extent those characters' behaviours, in the Greek legend of the two bears.

The Bears' story

According to Greek mythology, the bears Ursa Major and Ursa Minor found their way into the night sky because of one of Zeus' many sexual indiscretions. Zeus (known as Jupiter to the Romans) was chief of all the twelve gods who lived on Mount Olympus, most of whom were related to one another. Among them were Hera, Zeus' sister and wife, and Artemis, his illegitimate daughter. Hera was the goddess of marriage, Artemis the goddess of the wilderness, wild animals and the hunt. Armed with a bow and arrow and accompanied by her loyal band of nymphs, Artemis would hunt (and also, confusingly, protect) lions, panthers and stags in the mountain forests and surrounding lands. Nymphs in Greek mythology are always female. They feature quite heavily in the neoclassical paintings of the eighteenth and nineteenth centuries, where they are generally depicted as beautiful and naked. They were associated with wildness and nature, often personifying a specific natural feature. For example, Callisto, one of Artemis' nymphs and a central figure in this story, was a wood nymph. Because of these associations, and perhaps because of her father's reputation, as a sign of their loyalty to her Artemis required the nymphs to make a vow of chastity.

Naturally, this didn't stop Zeus, who came to Callisto in disguise (some say as Artemis) and seduced her (or raped her, depending on whose interpretation you read). When Callisto gave birth to his son she named him Arcas. Zeus then turned her into a bear, protecting her against the jealousy of Hera and the anger of Artemis.

While Callisto was a bear, Arcas grew up. He became a master with the bow and arrow and a keen huntsman. One day, out hunting in the forest, he spotted a bear and took aim. That bear turned out to be his mother, Callisto. Zeus again stepped in, this time turning Arcas into a bear so that he might recognize his mother and so saved Callisto's life. He then transported them both into the heavens where they became Ursa Major (the Great Bear) and Ursa Minor (the Little Bear).

However, the story was not quite over for the bears. Hera found out about Callisto. Choosing to punish her rather than Zeus, she pleaded with Poseidon (god of the sea and brother of both Zeus and Hera) never to allow the bears to bathe in his immortal waters. As a result, so the story goes, the two constellations never dip below the horizon. They are always visible in the night's sky, for most of the northern hemisphere at least. Boötes, the herdsman, accompanied by his hunting dogs Canes Venatici, keeps the bears together and away from the sea on their journey around the pole.

John Bevis and his Atlas Celeste

The image overleaf of Ursa Major comes from a beautiful set of star charts produced by John Bevis in the mid-eighteenth century. Like most star charts these were produced to entice as much as to inform. They were produced at great expense, as the title page announces:

> The Expense of the Engravings was immense, as the most Capital Artists in Europe were employed in executing them, and the learned and ingenious Delineators and Directors of the Work had determined to sell it by way of Subscription at Five Guineas the Set. The heavy Charge attending it, rendered some of them Insolvent, others were removed by Death, which with divers adverse Occurrences were the Means of retarding the Publication until the

The constellation Ursa Major

present Period 1786. Many of the Copies have been destroyed by Fire and Removals; the few that remain are now offered at One Guinea and a Half each Set.

This elegant and useful Work is not, nor ever had been in the Hands of any Bookseller. The Copies saved are all of the first Impression, and will be an Ornament to any Library, and highly worthy of the Notice and Patronage of every Lover of the Sciences.

The emphasis on 'ornament' is interesting because it gives an idea of how Bevis was expecting the charts to be used. They were not only for serious astronomers with an all-consuming passion and dedication to the subject; they were designed also for the fashionable elite, who accumulated attractive and well-stocked libraries in order to appear erudite. Astronomy was a particularly popular way of showing off.

But fashions change. There seem to have followed years of neglect in which this remarkable early atlas was almost entirely forgotten, until in 1997 when the Manchester Astronomical Society (MAS) took it upon itself to track down all the remaining copies from around the world to promote awareness of the charts and their creator.

John Bevis was an English doctor and astronomer best known for his atlas and his discovery in 1731 of the Crab nebula, sometimes called M1, in the constellation Taurus. He is described as its discoverer (or rediscoverer, since Chinese astronomers noted its existence back in 1054), because he marked it in this atlas.

Bevis was one of England's early grand amateurs. He earned his living as a doctor while devoting his spare time and spare income to the pursuit of astronomy. In the eighteenth century England had precisely one professional astronomer, the Astronomer Royal, working at Greenwich; so to talk about amateur and professional astronomy in the modern sense has little meaning. Professional astronomy during this time meant the type of practical astronomy needed for navigation: it was essentially dull, monotonous work that no one would really choose as a hobby. Amateurs, meanwhile, did the fun stuff. They discovered comets and, towards the turn of the century, even new planets and asteroids. These amateurs were often doctors or clergymen, with a good education, a reliable income and sufficient spare time to pursue astronomy as a pastime. Grand amateurs like Bevis could boast telescopes that might rival in power and size those bought by the king for his astronomers, something amateurs today can only dream of. It was in his own private observatory in Stoke Newington, north London, that Bevis discovered the Crab nebula, created his atlas and was one of only two British astronomers to observe and so confirm Edmund Halley's prediction of the return of his comet in 1759. He died a heroic astronomical death at the age of seventy-six, by falling from his telescope.

Each plate in Bevis's atlas is dedicated to a different contemporary patron of astronomy. The Ursa Major plate is dedicated to the University of Cambridge, which supported Isaac Newton throughout his career. Newton's work, and the subsequent popular interest in trying to understand it, gave rise in the eighteenth century to a whole astronomy industry. Following Newton's death, numerous textbooks were published claiming to explain Newton's work, but without the maths. New careers opened up for those with adequate education and understanding of astronomical matters, as the demand grew for travelling lecturers. These lectures took place in various locations, one popular venue being the coffeehouse, sometimes termed the Penny University for its role in providing lecture courses for all at a penny a time. This fashion, typical of the eighteenth century, was driven by the aristocracy and monarchy. The royal court took an interest in astronomy, especially after the arrival of George III in 1761, and gradually this filtered down to the lower, aspirational classes. Bevis, in the pricing of his atlas, the dedications he gave each plate and in his suggestion that it would make a great 'ornament' to any library, was very much placing his work at the top end of this hierarchy.

The tilting Earth and why we have seasons

A central theme in the Greek myth about the Bears is the apparent movement of the heavens. As we have seen, according to this mythology, the Bears never dip below the horizon — and indeed, for all observers living north of Beijing, New York and Athens they don't, at least the most familiar parts of them don't. For people living farther south they do dip, and as you get even further south, they disappear altogether. For observers living south of Alice Springs, Rio de Janeiro and the Kalahari Desert, Ursa Major and Ursa Minor never rise. Why?

The Earth, as we know, turns on its axis, making the Sun (in the day) and the stars (at night) appear to rise and set. This has been well known to us since around 400 BCE. The spinning Earth orbits the Sun, in roughly the same plane as all the other planets in our solar system. If looked at from a very long way away, our solar system would look like a rotating flat disc, with the Sun at its centre. It is in relation to this flat plane that we say the Earth tilts. What we mean is that its axis, the imaginary rod joining the North and South Pole around which the Earth spins, is at an angle to the plane of the Earth's orbit around the Sun. Measurement of the exact angle of this tilt, pleasingly termed the 'obliquity of the ecliptic', has been the object of a good deal of painstaking research and gradual refinement over the millennia. The ancient Greeks gave it a figure of between 23° and 24°, and this is still pretty much what we use today.

The tilt of the Earth means that the pole star, Polaris, never sets for anyone north of the equator. As we've seen, it also means that Ursa Major and Ursa Minor never set for anyone north of Athens. As we move farther south there are fewer and fewer northern

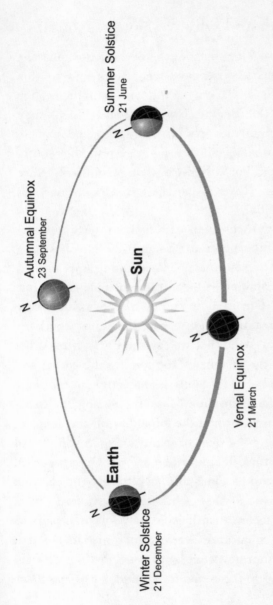

Summer Solstice
21 June

Autumnal Equinox
23 September

Sun

Earth

Winter Solstice
21 December

Vernal Equinox
21 March

The Sun, the Earth and the seasons. This diagram shows how the Earth is tilted in relation to its orbit and how it is aligned at different times of the year.

hemisphere stars which are always visible. When we reach the equator the balance shifts: from the equator southwards it is the southern hemisphere's constellations that we can mainly see and, as we get closer to the South Pole, more and more of these appear never to set.

The tilt of the Earth is important for another reason: it's why we have seasons.

In the northern hemisphere, it is winter when the North Pole points away from the Sun (in fact, it points so far away that for several days at the pole the Sun never rises). During this time, the South Pole is pointing towards the Sun; so the southern hemisphere has more hours of sunlight and gets warmer – it's summer. Conversely, summer in the northern hemisphere occurs when the South Pole points away from the Sun. This time the northern hemisphere has the most daylight and therefore gets warmer. The point at which the Earth's axis is exactly in line with the Sun, pointing either directly towards or directly away from it depending on your position, is known as the solstice. This is when it's the longest day for half the world and the shortest day for the other. The northern hemisphere's winter solstice (summer solstice for the southern hemisphere) falls around 21 December, while the northern hemisphere's summer solstice (winter solstice if you're in the south) falls around 21 June.

These and other astronomical markers are used in almost all calendars around the world, today and historically. In autumn and spring the Earth's pole neither points towards or away from the Sun. At this point, when the tilt is in line with the path of the Earth's orbit, day and night are of equal length. This is the equinox. The vernal equinox is sometimes considered to be the beginning of spring, falling around 21 March in the northern hemisphere and 21 September in the southern, and the autumnal equinox to be the beginning of autumn. Easter in the Christian calendar is calculated from the northern hemisphere's vernal equinox, as is Passover in the Jewish calendar. In many much earlier civilizations the vernal equinox was also celebrated, often as the start of the year, marking the end of winter, and bringing renewal and new life.

Stonehenge on Salisbury Plain in Wiltshire, England, is probably the most famous remaining evidence we have of the fact that these astronomical markers were celebrated in ancient cultures. There are many interpretations of what, exactly, Stonehenge was for, and the debate remains heated. I don't, however, think it's too controversial to say that Stonehenge, dating from around 2000 BCE, is aligned towards the summer solstice sunrise in one direction and the winter solstice sunset in the other. A more recent ancient monument to celebrate these astronomical markers is Machu Picchu in Peru, where a column of stone marks the equinoxes. At exactly midday on each equinox the Sun appears to sit on top of this column, the Intihuatana.

The Greeks

Although the four northern hemisphere constellations – Ursa Major, Ursa Minor, Boötes and Canes Venatici – can be brought together to tell a single story, they did not all originate from the same point in history. The earliest surviving written mention of Boötes is in Homer's Odyssey, the story of Odysseus' adventures as he returns home to Greece after the Trojan Wars, written over 2,700 years ago. Homer describes Odysseus using the constellation Boötes as a guide as he sets sail from the island of the nymph Calypso.

Homer, like his near contemporary Hesiod, is thought to have compiled and recorded stories that had already been passed down through the generations by word of mouth. By 700 BCE, the ancient Greek empire incorporated much of southern Europe, the Middle East and North Africa, but trade and communication stretched to countries much farther afield. Gradually, various cultures' oral mythologies were incorporated into the Greek mix. Our knowledge of their precise origins may be patchy, but the constellations we think of and often describe as Greek or Ptolemaic (after the Greek astronomer who wrote them all down) are often nothing of the sort. When it comes to the newer constellations, like Boötes's hunting dogs, Canes Venatici, it is much easier to trace origins and history.

The Heveliuses

The constellation Canes Venatici was first described in print in 1690, by a husband and wife team, Johann and Elizabeth Hevelius. Johann Hevelius was born in Danzig (Gdansk) in Poland in 1611. He studied Polish, mathematics, astronomy and instrument making at the local gymnasium (school) and with private tutors, and in 1630 went away to study law at the University of Leiden, where he learned more mathematics and astronomy. A year later he travelled to London, then Paris, calling on various men of science along the way. He returned home in 1635 to marry and to begin work in his father's brewery. Using the money he made in his professional life as a brewing merchant and property owner Hevelius built himself a well-equipped observatory on the roof of his house.

Elizabeth Hevelius (née Catherina Elizabetha Koopman) was Hevelius's second wife. She was an intelligent sixteen-year-old when his first wife died, and well aware of the restrictions women faced in pursuing science. At that time, women could neither go to university, join any academic societies nor publish papers in academic journals; although they could publish books in their own name, they would be heavily criticized and ridiculed if they were to present too much of the material as the result of their own independent research. As late as 1740, Émilie du Châtelet, best known today as Voltaire's mistress, wrote her own treatise *Institutions de Physiques* (Lessons in Physics). Though well respected among her immediate contemporaries she still took the precaution to state in her introduction that the book was written for mothers, like her, to help them teach their sons, rather than as a result of her own new research.

Little is known of Elizabeth before her marriage to Hevelius except that she was a bright girl from a rich merchant family. As the wife of a respected astronomer, she knew she would have her own private tutor and the opportunity to participate in world-class research. This is not quite as devious as it sounds. Marriage at this time was often primarily a business relationship linking families in rival businesses to consolidate profits. It was also relatively common for wives to share in the work of their husbands, whether they received credit for it or not. What is unusual about the Hevelius case is that Elizabeth's work was recognized.

They married in 1663. In 1679 Hevelius's observatory, including his instruments and many of his notes, was destroyed by fire. Work on a new star atlas ground to a halt. But with the help of the King of Poland, Jan III Sobieski, they were able to rebuild the observatory. The Heveliuses, like Galileo, knew the value of elaborate displays of gratitude. Galileo had named the four moons of

Jupiter (which he discovered with his telescope) the Medici Planets, in honour of his patrons. To show their thanks, first Johannes named one of his new constellations the 'Shield of Sobieski' or Scutum for short, then Elizabeth, in her introduction to the atlas, sang the king's praises. She drew specific attention to her husband's constellation and diplomatically entitled the atlas *Prodromus Astronomae Fermamentum Sobiescanum*.

Johannes died before the atlas was complete, but Elizabeth continued to work on it. The catalogue, complete with their new constellations Scutum and Canes Venatici, was published in 1690.

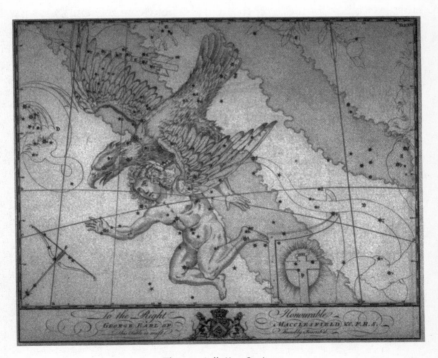

The constellation Scutum

Both constellations have stood the test of time, becoming two of the eighty-eight constellations the international community of astronomers officially recognized in 1930. This image of Scutum, again taken from Bevis's atlas, shows Hevelius's southern hemisphere constellation, Scutum in the bottom right-hand corner as a crucifix on a shield, lying almost entirely within the band marked out as the Milky Way.

Incidentally, one 'deep sky' object to be found in the constellation Scutum that you might have heard of, but cannot quite see with the naked eye, is the Wild Duck Cluster. With an apparent magnitude of 6.3 it should be visible through a pair of binoculars. It is an open cluster (we'll meet more of these in the next chapter) and gains its name from the supposed similarity in appearance to a flock of wild ducks flying across the sky.

The main constellation shown in Bevis's illustration is Aquila, the eagle depicted carrying a young man, Ganymede, to Zeus to act as the gods' new wine-bearer. Aquila, too, we will meet in more detail in the next chapter.

Crux

For observers south of the equator, the bears and their story are of limited practical interest. The southernmost constellations we know today almost all date from the sixteenth, seventeenth and eighteenth centuries and were invented by European travellers and mapmakers, who largely ignored the constellations in use in different parts of the southern hemisphere. The most familiar of today's southern constellations is the Crux, or Southern Cross, which is simply four bright stars that can be joined together to make a cross. You'll find it a little way south of the centre on spring's star chart, which means it's nearly but not quite overhead in the April night sky. Unlike the northern sky, there is no single pole star for the southern hemisphere, but there are constellations close to the point in space directly above the South Pole that go round this point. These are circumpolar constellations, circling the celestial pole. The Southern Cross is a circumpolar constellation. The constellations we met in the northern hemisphere that circle the pole star Polaris and appear never to set for much of the northern hemisphere are similarly circumpolar constellations. Crux never sets for much of the southern hemisphere, at least for those living south of Sydney, Montevideo and Cape Town. It can be easily seen, too, in much of the northern hemisphere – as far up as Hong Kong, Mecca and Havana.

The Southern Cross was known to ancient Greek astronomers, who could see it even from places as far south as Egypt, but in those days these four bright stars were regarded as part of the constellation Centaurus. The effects of precession – the wobble of the Earth on its axis – meant that gradually the stars known as the Southern Cross fell out of view for northern astronomers. So when European explorers travelled south in the sixteenth century they thought these four bright stars were a new discovery. Considering them a good omen, they renamed the group Crux, or the Southern Cross, depicting the new constellation as a Christian crucifix. You can see it depicted in John Bevis's atlas on the following page.

Double stars

Many of the stars we see in the night sky are more than just single stars. They can be double stars, nebulae, star clusters or even whole galaxies. To make these distinctions we would generally

APRIL

To COUNT CYRILLUS
Hetmann, and Chamberlain in Ordinary
President of the Imperial Academy of Sciences at St. PETERSBURG,
This Table is

DE RASUMOWSKY
to The EMPRESS of RUSSIA &c.&c.&c.
and Fellow of the Royal Society of LONDON,
most humbly inscrib'd

The constellation Crux

need at least a small telescope. To see the kind of beautiful, colourful images we're used to seeing in the press announcing astronomers' new discoveries we would need giant telescopes, possibly even located outside the Earth's atmosphere. We would also need a range of apparatus to interpret the information those telescopes produce and turn it into attractive pictures. Occasionally, however, we can just about make out some of these 'deep sky objects' with the naked eye. Some of the easiest to make out are double stars. Ursa Major and Crux are particularly good sources for these.

As the name would suggest, double stars are stars which, on closer inspection, appear to be not one star but two. There are two types of double stars: visual binaries and optical binaries. Visual binaries are pairs of stars that revolve around each other (or interact in some other way) due to the gravitational pull exacted on one by the other. Optical binaries are stars that appear from Earth to be close together but are actually light years apart. The first visual binary, or true binary, star to be discovered was Mizar in Ursa Major. The Jesuit Italian astronomer Giovanni Battista Riccioli discovered this was a true binary in 1650. By looking at this star with a telescope, Mizar was found to be two stars revolving around one another. But Mizar was considered a double star long before 1650 because of its closeness to a neighbouring star, Alcor. The beauty of this double for stargazers is that it is visible, just about, without a telescope. If you first go back to the spring star chart, then the sky, and find the saucepan shape of Ursa Major, the second star from the end of the handle is Mizar. With good eyesight it is just about possible to see the slightly fainter Alcor very close to Mizar. This pair, known to the Arabs as 'horse and rider', was traditionally used as an eye test – if you could make out Alcor your eyesight was very good.

Canes Venatici provides another example of a double star that is worth a look, in the form of the rather romantically named Cor Caroli or Charles's Heart. Go back to the illustration of Ursa Major from Bevis's atlas on page 22 and you'll see it clearly marked. Yapping at the Bear's back legs are the heads of Boötes's hunting dogs, Canes Venatici. Between them is Cor Caroli, shown as a heart and crown. This double star is made up of two blue-white stars, impossible to make out with the naked eye, but with binoculars you can see them as two separate stars.

The star's full name was Cor Caroli Regis Martyris, in memory of King Charles I, executed during the English Civil War in 1649. It was named in 1725 by Edmund Halley, the astronomer best known for his prediction of a comet's return. He had reason to thank the royal family when Charles I's son, Charles II, was restored to the throne and supported the foundation of both the Royal Society and the Royal Observatory in Greenwich, a boon for Halley and his fellows.

Crux, too, has an excellent bright double star to look out for. Like the pair in Ursa Major, Crux's brightest star, α Crux, is a visual or true binary. Like Cor Caroli, it needs some magnification to resolve it into its constituent parts. To the naked eye it looks like a single, albeit very bright, star.

The stars we have met so far have formed old and new constellations, with histories varying from the obscure to the well documented. The bears and the cross provide a good starting point, since they are familiar, but much of the sky reveals some real surprises.

CHAPTER TWO

May and Hercules

THE TWELVE LABOURS of Hercules is probably one of the most familiar of the Greek myths. (The Greeks knew him as Heracles, but the Roman version now seems more familiar to us and is the one used in astronomy; so I'm using Hercules throughout.) The constellations in this chapter are in many cases much less well known, and in some instances harder to see than those in Chapter 1, but they do start to show how different parts of the night sky link up.

The Greek myth tells how Zeus' illegitimate son Hercules was driven mad by a furious, jealous, vengeful Hera, Zeus' wife. In his madness he killed his wife and children. When he realized what he'd done, he consulted the Oracle at Delphi who told him to go to King Eurystheus and do whatever he said. King Eurystheus set him twelve seemingly impossible tasks, which, if successfully completed, would rid him of his guilt and make him immortal.

We see Hercules' success commemorated on vases, in paintings and in sculpture from antiquity to recent times.

Coffee and cartoons

The Disney film *Hercules*, which came out in 1997, also helped to keep many of the characters, if not the finer details of the story, alive to modern audiences. The film is, as you might expect, a much sanitized version of the Greek original. Hera is Hercules' doting mother rather than the vengeful wife of his father. The hero's motivation for carrying out his labours is changed, too: rather than acting out of guilt for killing his wife, he instead acts from an anachronistic desire to be reunited with his biological family. Much is made of Hercules being Zeus' son, as though this were something unusual, glossing over the fact that almost all the gods and mortals in Greek mythology are siblings, conquests or children of Zeus. The tasks, too, are simplified. Only the killings of monsters such as the multi-headed hydra are shown. However, a positive feature of the film is that it teaches familiarity with the characters, their names and basic traits. It's great for occupying small children first thing in the morning, as you have your coffee and recover from a night of stargazing. Pleasingly, the film even ends with Hercules being created out of stars in the night sky, thereby demonstrating the correlation between the myth and the constellation.

The idea of becoming familiar with ideas out of context and then having their true meaning revealed is still often frowned upon, particularly among scientists, but it can be an effective and effortless way to crack a new subject. I can still remember being delighted in a high school physics lesson to learn what Brownian motion really was after years of knowing the term from *The Hitchhiker's Guide to the Galaxy*. For those of you with a dimmer memory of that particular lesson in school, Brownian motion is the random motion of particles in a fluid which has only the slightest connection with a hot cup of tea. The fact that I already knew the term made the actual concepts much more memorable.

This method is exploited a good deal in museums where it's common practice to use a familiar historical name as a hook to draw people into an exhibition. The exhibition may then try to challenge commonly held beliefs about that person, or add new insights, but that basic familiarity is invaluable as a means of giving the audience a way in. This can work just as well with science, even though it's often deemed trivial or unscientific. For example, most people know their star sign and probably know the name to be attached to a constellation. Those born under the Sun sign Leo, for example, are born at the time of year in which the Sun is found in the constella-tion Leo. If it wasn't for the fact that astronomers don't like to admit they have anything to do with astrology, this could be a useful way of helping people build up their knowledge. The same prin-ciple works here, that by knowing the Hercules story, even if it's just the Disney version, you are better placed to understand and remember the constellations associated with that story than if you came to them cold.

The labours of Hercules

Hercules' twelve labours begin with the killing of the Nemean lion and returning to King Eurystheus with its skin. Unbeknown to Hercules, the lion was immortal and could not be killed with arrows. However, Hercules managed to overpower him by choking him to death with his bare hands. References to this part of the story are found twice in the night sky, as the constellation Leo the lion and in the constellation Hercules, where depictions always show him wearing the lion skin.

Hercules' second labour was to kill the multi-headed Lernean hydra, a monster whose heads would grow back each time they were severed. Taking advice from his nephew, Iolaus, Hercules learned he needed to scorch each stump as soon as he had cut off the head and so he successfully completed his second task. The Lernean hydra from the story is represented by the constellation Hydra, generally depicted with only the last of its nine heads still surviving.

Tasks three, four and five – to capture the golden stag of Artemis, the Erymanthian boar and to clean the Augean stables – do not have associated constellations. Task six, however, has three. In this task, Hercules was called upon to destroy the vicious, metallic, man-eating birds terrorizing the forests around Lake Stymphalia, otherwise known as the Stymphalian birds. He did this by scaring some away and shooting the rest down with his arrow (sometimes associated with constellation Sagitta). Curiously, the constellations associated with these birds are the very unmetallic eagle, swan and vulture. These constellations are also known as Aquila the eagle, Cygnus the swan and Lyra, today depicted as a lyre but previously depicted as a vulture holding a lyre. You can see the vulture version of Lyra in the illustration of Hercules below. Lyra is by his right arm and knee.

You'll notice Hercules is upside down. This is because, in the night sky, his feet are pointing towards the northern pole star, Polaris. But he is not always upside down to us, the observer. As the Earth turns, he, like Ursa Major, Ursa Minor and all the other stars close to the pole (the circumpolar stars), appears to go round the pole star. This means he effectively points in all directions, depending on the time of year. John Bevis was following the convention, established in terrestrial cartography, that put north at the

The constellation Hercules

top of the page. So Hercules appears upside down.

Labours seven, eight, nine, ten and twelve – to capture a bull on Crete and the wild horses belonging to the god Diomedes; to take the belt of Hippolyte, Queen of the Amazons; to get past a two-headed dog and aggressive huntsman to collect some cattle; and to capture Cerberus, the three-headed dog guarding the gates of the underworld – are, like three, four and five, not represented in the night sky, unlike task eleven, which is. In this labour, in order to steal the golden apples of the Hesperides, Hercules has to get past Draco the dragon (sometimes called Ladon).

As the plate of Draco shows quite neatly, the dragon's head can be found just below Hercules' foot, while his body is wrapped round the pole star, just skimming the feet of Ursa Minor. Incidentally, the circle on this plate is not a circle round the pole star, showing the circumpolar stars, but a circle around what would be the North Pole, if the ecliptic or zodiac constellations were parallel with our equator. In other words, Bevis has untilted the Earth. His co-ordinates are given purely in terms of the stars and how they relate to each other rather than how they appear from Earth. This is a tradition he adopts from the ancient Greeks, and

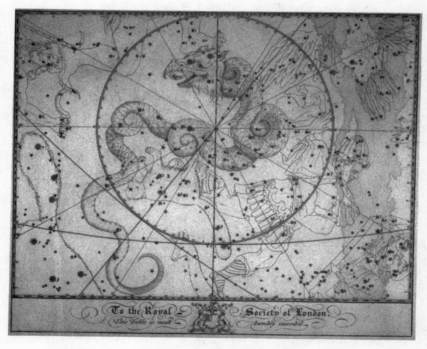

The constellation Draco

specifically Ptolemy, on whose surviving works much subsequent knowledge of Greek astronomy is based. It is a system Bevis uses throughout his atlas, but it's only here that it could become potentially confusing.

To find these various constellations associated with the Hercules story look at the star chart for spring. You will find these constellations in two basic groups, one on either side of the constellations Boötes and Ursa Major we met in the last chapter. Hercules, Draco and the three birds are on one side; Leo and Hydra are on the other.

The constellations Leo and Hydra and those around them should be easily visible in May from both the northern and southern hemispheres, so we'll start with them. Leo sits near the ecliptic – an imaginary line in the sky along which it was once believed the Sun travelled. Now, of course, we know that we orbit the Sun rather than the other way around, but the line remains as a reference point, as does the band of sky on either side of it. This imaginary line circles the Earth at an angle of around 23° to the equator (this is the obliquity of the ecliptic) and is parallel with the plane of the solar system, which means it is where we find all the celestial bodies in our solar system – the Sun, the Moon and all the planets. This is why the zodiac constellations have always been considered so important. They are not as a rule especially bright or impressive constellations, but they are the constellations in which we find these bodies. The Sun's position within these constellations at different times of year provides an easy way of marking the seasons, while the planets' movements through these constellations were seen as messages from the gods.

Leo's bright star Regulus should help you find the rest of the constellation, though many of the other stars in Leo are quite faint to the naked eye. Lying underneath Leo, and extending along the ecliptic below Virgo all the way to Libra, is the constellation Hydra, the largest constellation in the night sky.

Occultations

Regulus is a very bright star, not easily missed, which makes its 'occultations' by the Moon, and occasionally by a planet, interesting and possible to watch. Regulus is found in the ecliptic so our view of it is sometimes obscured by the Moon or one of the planets. This is called an occultation; we say Regulus has been occulted by the Moon, say, or by the planet Venus. In fact Regulus is often occulted by the Moon, roughly once every twenty-eight days, and the occultation can last for about an hour, though where you can see it from varies each time. Occultations by planets are much less frequent. The last occultation of Regulus by a planet was by Venus in 1959; the next will again be by Venus, but not until 2044.

Regulus is not the only star we can see with the naked eye occulted by the Moon or one of the planets, though it is the brightest. Planets, too, are sometimes occulted by the Moon or another planet. To see when the next occultation is due to occur, and where you will need to be to be able to see it, look for listings in monthly astronomy magazines or on the International Occultation Timing Association (IOTA)'s website.

The instruments of astronomy

Combining some of the faint stars in the area of the sky just above Hydra and below Leo, Hevelius (see pages 29–31) created the constellation Sextans. This was named after his observing instrument, the sextant, meaning a sixth of a circle, destroyed when his observatory burnt down in 1679.

Astronomical sextants and quadrants (a quarter of a circle) were used by astronomers before the invention of the telescope, and then later adapted by simply putting a telescope in place of the 'sights'. The sights, like gun sights, would typically be two pin holes set along one of the radii of the sixth or quarter circle. Once the star and both sights were in line the astronomer could then measure an angle along the curved edge of the circle. This gave one co-ordinate; time gave the other needed for plotting or finding a given star on a star map. Alternatively, a star's position would be measured in terms of its distance from another star. Modified versions of these sextants and quadrants were some of the earliest types of astronomical telescope. These instruments are important to our story as they enabled the creation of the star maps that tell us so much about the constellations.

Sextans, near Hydra's head, is a small constellation made up of faint stars. Hydra has only one bright star (towards its head), making this constellation quite difficult to make out, too. Success comes from practice and becoming familiar with the shape the stars should be mapping out. The key to finding this constellation is to look from an appropriately dark spot, and to keep looking with a good idea of what you're looking for. The same is true of a lot of stargazing, especially once you start looking for the more obscure constellations.

The famous eighteenth-century astronomer Caroline Herschel put this rather well. She had been pushed into stargazing by her older brother, but, determined to make the best of it, had practised as often as she could, keeping a journal as she did so. As she described later in life:

I began Aug 22, 1782 [to practise stargazing] ... but it was not till the last two months of the same year before I felt the least encouragement for spending the starlight nights on a grass-plot covered by dew or hoar frost without a human being near enough to be within call.

In other words, it took her about three months of regular practice to be familiar enough with the night sky to enjoy her new 'hobby'. In fact, she was being trained to become her brother's assistant. He had shot to fame with the discovery of the planet Uranus, and was now working on a detailed survey of all the nebulae, star clusters and double stars he could see with his telescope. This meant she needed an appropriately detailed understanding of the heavens. She also needed to know all the constellations visible from the northern hemisphere and exactly how to use her telescope. Her experience goes to show that you shouldn't be disheartened if you struggle to find some of the fainter constellations – even after practice.

Nebulae and star clusters

For his second labour, Hercules is sent to kill the hydra of the swamps of Lerna (a nine-headed serpent, though only one of the heads is immortal, and only one is represented in the constellation). While he is busy fighting the serpent, Hera sends a crab, the constellation Cancer, to hinder his progress. When Hercules hardly notices the crab grabbing his toes and steps on it in the course of the fight, the plan fails. To show her gratitude for Cancer's efforts Hera places the crab in the heavens. You'll find Cancer in the night sky a little further round the ecliptic, next to Leo and just above hydra's head.

Cancer is best known to astronomers as the home of deep sky object M44, otherwise named the Beehive cluster. This is an 'open cluster' just about visible, in good conditions, as a smudge to the naked eye. Up close, with the aid of a telescope or binoculars, it might look a bit like a beehive (hence its name); to the ancient Greeks and Romans it evidently looked more like a manger since they named it Praesepe (Latin for manger). According to the Greek myth associated with this group of stars, Praesepe was the manger out of which the donkeys of gods Dionysos and Silenus ate before they rode into battle with the Titans (Zeus' parents, uncles and aunts). The two donkeys are the stars on either side of this cluster, γ-Cancri and δ-Cancri, also known as Asellus Borealis and Asellus Australis, the northern and southern donkey respectively. Another cluster found in this month's constellations is Great Hercules, a globular cluster, otherwise known as M13. This is also visible to the naked eye as a blurred object rather than a sharp, single point of light.

As the name suggests, clusters are groups of stars, held together by gravity, which all travel with one another. Globular clusters are particularly tight groups of old stars that appear spherical and orbit a galaxy. Open clusters, such as the Beehive cluster, are much less tightly grouped and are made up of much younger stars. The stars in a globular cluster are generally all about the same age, and so at the same stage in their lifecycle. Stars in any globular cluster are usually past the main sequence stage of their evolution or lifecycle, which means they have finished turning hydrogen into helium at their cores and have started turning helium into carbon or even carbon into heavier elements. Because of this you can give a definite age to the whole cluster. The estimated age of M13 is eleven thousand million years old, (1.1×10^{10} years old), making

The Hercules globular cluster, M13

it one of the oldest objects in the universe. Globular clusters comprise a few hundred thousand stars but no gas or dust, so the assumption is that this has long ago been turned into stars. We know they orbit galaxies, but no one quite knows how they form or what their relationship is to the formation of galaxies. We do know that clusters do not always orbit the galaxy they formed with. Galaxies can sometimes give their globular clusters to their neighbours. Right now our neighbouring galaxies, Sagittarius dwarf and Canis Major dwarf, are in the process of donating some of their globular clusters to our own galaxy, the Milky Way.

Stars in open clusters, are typically less than a few hundred million years old. They have only ever been found in galaxies where star formation is still going on. They comprise a few hundred stars, formed at the same time from the same molecular cloud. A molecular cloud is cloud molecular hydrogen. It is the star-forming part of a nebula. It doesn't produce any light itself, though can sometimes be seen if light from another source is reflected by some of the particles it contains. This is similar to the Moon and the planets in our solar system. These bodies don't generate their own light, but we can see them because they reflect the light from the Sun.

The most well-known open cluster, and one of the few that can be resolved into its constituent stars (or at least some of them) with the naked eye, is the Pleiades in the constellation Taurus. We will come back to the Pleiades in later chapters. The Beehive cluster in Cancer represents a more typical observing experience.

While the stars in a globular cluster all stick together until the bitter end, stars in open clusters will gradually disperse. This is why we can say open clusters contain only young stars. Once those stars begin to age, the stars spread out and stop forming a cluster. They are important in the study of stellar evolution because we know they all formed at the same time, from the same material, and are all roughly the same distance from Earth. This means any difference between two stars is attributable only to their difference in mass.

A long time ago, all night sky objects that looked like smudges were classified as nebulae (from the Latin for mist). Now the term refers only to a very specific group of objects, molecular clouds and newly forming stars. Some we see only as dark patches called dark nebulae in otherwise very bright areas of the sky. The Coalsack nebula in the constellation Crux in the Milky Way is a good example. This dark patch of sky, far from showing an absence of stars, is in fact a dense area of dust and gas that obscures our view of the stars. The nebulae we can see as bright objects are sometimes called emission or reflection nebulae, but mostly they're just called nebulae. One nebula we can see with the naked eye in the May sky, and in a constellation associated with the Hercules story, is NGC 7000 in Cygnus. It is visible as a hazy cloud if not to the naked eye then certainly with binoculars and is marked on the spring star chart just to one side of the bright star Deneb in Cygnus.

Nebulae and star clusters were first catalogued in the eighteenth century, when the fashion for comet hunting really took off, so that astronomers didn't mistake them for new comets. The French astronomer Charles Messier created a list of objects that, like comets, looked essentially like blurred stars viewed through a telescope, but weren't. He, like his fellow eighteenth-century astronomers, wasn't interested in what these objects were; he just wanted to make sure people knew where they were so that they didn't take them for comets. The nebulae and star clusters in his catalogue are still often referred to as Messier objects, and given names like M13, short for Messier 13. His catalogue was followed by Herschel's and later by J. L. E. Dreyer's New General Catalogue of nebulae and star clusters in the 1880s. The nebula in Cygnus has a New General Catalogue name, NGC 7000.

Related Greek myths

Besides being the multi-headed monster from the Hercules story, the constellation Hydra also appears in Greek mythology as a water snake brought to Artemis' twin brother, Apollo, by the crow Corvus. According to this story, Apollo sent Corvus with a cup, Crater, to fetch him some water from a spring so that he could make a sacrifice to his father, Zeus. On the way, however, the crow was distracted by some nearly ripe figs. He waited by the figs for them to ripen, ate them, then needed an excuse for why he'd taken so long. He took a water snake back to Apollo and claimed it had been blocking the spring. But Apollo, god of the Sun, patron of the arts and gifted with the art of prophecy, knew better. Refusing to accept this excuse he flung the crow, the cup and the snake into the sky where they became Corvus and Crater on the back of Hydra.

Corvus and Crater, although visible from the northern hemisphere in the May sky towards the horizon, are probably best seen from the southern hemisphere. The stars making up Corvus and Crater, as you can see from the image from Bevis's atlas, are

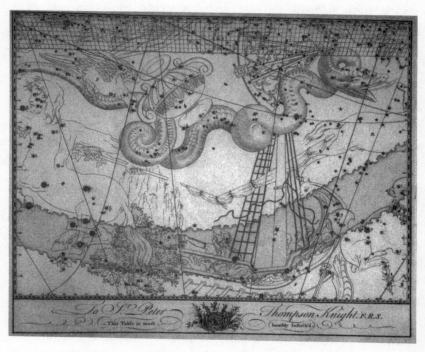

Hydra with Corvus and Crater on its back, and Sextans near its head

not the brightest stars in the sky but can nevertheless be seen with the naked eye in the right conditions. Corvus' brightest star is apparent magnitude 4. (The human eye can typically see stars only at apparent magnitudes of 6 and above.) The best way to locate Corvus' and Crater's stars is to look on a clear, dark night for the Southern Cross, Crux, which you can see at the bottom of this picture. Then work your way up, finding the bright star Alpha Centauri in the constellation Centaurus. Alpha Centauri is actually a group of stars, the third brightest star grouping in the night sky: it just appears as a single object to the unaided eye. Above this you should be able to find this region of dimmer stars, which, with a lot of perseverance, you can resolve into Corvus and Crater.

The constellation Centaurus is linked to several centaurs (creatures that are half-human, half-horse) in Greek mythology including a character in the Hercules story called Chiron. Unlike most centaurs in Greek mythology, Chiron was reserved, cultured and well read. He became the teacher of Hercules and Achilles and, according to some accounts, was killed accidentally by Hercules as he pursued one of his labours. In recognition of his work as a teacher, Zeus placed him in the heavens.

If we now move to the other side of the constellations Boötes and Ursa Major we come to Hercules and Draco. A little farther out, and closer to the horizon in the northern hemisphere's spring sky (and out of view for most of the southern hemisphere), we come to the three birds.

In Hercules' sixth labour he kills and drives away the Stymphalian birds with his bow and arrow. One of these arrows found its way into the sky between Aquila the eagle and Cygnus the swan, and became the constellation Sagitta. To find the three bird constellations Aquila, Cygnus and Lyra, the lyre-carrying vulture, you need to locate what astronomers today call the 'summer triangle'. This triangle of three bright stars is most visible in the summer, although it can also be seen in the northern hemisphere's spring sky. The stars are Altair in Aquila, Deneb in Cygnus and Vega in Lyra. They stand out from the other stars in that region of the sky and are easily visible in a clear sky. Vega appears the brightest, followed by Altair, then Deneb (even though Deneb produces the most light it's much, much further away). Vega is the second brightest star in the northern hemisphere after Arcturus in Boötes.

Twilight and the summer triangle

The stars of the summer triangle are the first visible each night in the northern hemisphere's summer sky. We can see them even before the sky is properly dark; this is as well since at the height of summer, for a large part of the northern hemisphere, the sky never goes properly dark. We call this period of complete darkness astronomical twilight.

Most of us think of twilight as the period after the Sun has set and before the sky is completely dark. But astronomers categorize twilight into three different types, civil, nautical and astronomical. Civil is the lightest, meaning the point at which street lights and car headlights must go on. Nautical refers to a slightly darker period when the brightest stars, those traditionally used for navigation, are visible. Astronomical twilight is when the sky is completely dark. At the height of summer many places have whole nights without astronomical twilight, even places as relatively far away from the poles as London.

Vega is the brightest of the three stars that make up the summer triangle. It was also, like Thuban in Draco, an ancient pole star. Today Polaris is our pole star. As the Earth has wobbled slowly over thousands of years the site of the pole star has changed. Thuban was the star by which many of the ancient Egyptian pyramids were aligned, shining directly down their entrances. Vega is significant today as the zero point for apparent magnitude measurements.

Altair, the second brightest, is a main sequence star like our Sun and Vega, busy converting hydrogen to helium at its core by nuclear fusion. It is one of the brightest stars in the night sky and also, at only seventeen light years away, one of the closest. A light year is a measure of how long it takes the light from a star to reach the Earth. When we look at Altair, what we're actually seeing is what Altair looked like seventeen years ago. Since light travels at approximately 3×10^8 m/s (metres per second), a simple sum tells us that Altair, one of our closest stars, is approximately 160 billion (million million) kilometres away.

Deneb, the third star in the triangle, is a white supergiant and one of the genuinely brightest stars we know of. While its apparent magnitude, its brightness as it appears from Earth, is only 1.25, its absolute magnitude is −8.5. This makes it about 60,000 times as bright as our Sun. The absolute magnitude of a star is a measure not of how bright that star appears, but of how bright that star actually is. Our Sun has an absolute magnitude of around 4.8. Deneb is a very large star, thought to be around twenty times as massive as our Sun, and is consequently expected to have a relatively short lifespan in astronomical terms. In a few million years it will probably become a supernova.

Each of the birds associated with the summer triangle, Aquila, Cygnus and Lyra, also has an alternative Greek myth associated with it which better explains its particular form. Where Corvus the crow was sent on errands for Apollo, Aquila the eagle performed a similar service for Zeus. Most famously, he recruited a new wine-bearer for the gods in the form of Ganymede, the most beautiful boy he could find. The gods were delighted with Ganymede and decided to keep him. Ganymede sometimes features as part of the depiction of the Aquila constellation but he also appears on his own as Jupiter's largest Moon, one of the four discovered by Galileo. The other three Galilean moons are also named after Zeus' conquests: Callisto (whom we met in Chapter 1), Io and Europa.

Cygnus the swan has two alternative myths. In one, he is Zeus in disguise, out to seduce Leda, the Queen of Sparta. In another he is the swan friend of Phaethon, Apollo's mortal son. When Phaethon borrowed his father's sun chariot and crashed into the river, Cygnus dived down repeatedly, desperately searching for Phaethon. Pitying him, Zeus decided to place Phaethon's grief-stricken friend up in the night sky as Cygnus.

Lyra's alternative mythology is a little more complicated. The bird part of this constellation appears to be solely associated with Hercules' Stymphalian birds. The musical instrument, the lyre, seems to come from a totally separate story concerning Apollo. Apollo gave the instrument to Orpheus, his son, and one of Jason's Argonauts. With it he was able to drown out the sirens' song, thus saving his shipmates – Jason and other Argonauts – on the *Argo Navis*. With his beautiful playing he was also able to use the lyre to persuade Hades, brother of Zeus, Poseidon and Hera and god of the underworld, to release his dead wife Eurydice from the underworld (though he later lost her due to his failing to keep to the letter of the agreement).

These three bright stars, the summer triangle, along with Boötes and a few others were used by ancient astronomers and navigators as signposts in the sky. The fact that the stars you can see at any given time determine where you are on Earth has been a major reason for astronomy's importance over the ages. It means that if you know the time you can work out your location, and vice versa. To use the stars in this way, however, you need know only the positions of a few — and the brightest make the most obvious choice. These bright stars, among them Altair, Deneb and Vega from the summer triangle, were singled out for a very popular early astronomical instrument called the astrolabe. Until the invention of the telescope, this was the archetypal astronomical instrument, as this sixteenth-century painting of a Turkish observatory demonstrates.

The astrolabe in this painting is the circular object held by the third man from the right at the back. It comprises several plates. One is a grid marking out the brightest stars as their positions relate to one another. It also contains a circle showing the position of the ecliptic in relation to the given bright stars. This turns over a plate for a particular latitude (often described as a particular city) showing the observer's co-ordinates — where the horizon is, the zenith (a point directly overhead), and everything between the two. There is then a scale around the edge so that the grid can be turned to show which stars will be at the horizon, which directly overhead, and all those in between, at any time from that city. It can be used for telling the time by the stars or for showing what is where in the sky above you at any one time — what has just set, and what will rise soon. This information is useful to astronomers; it is also useful to astrologers casting horoscopes.

This painting, dating from 1581, shows astronomers at work with a variety of typical contemporary instruments at Taqi al-Din's observatory at Galata, Istanbul.

A couple of small astronomical quadrants, with sights rather than telescopes since the picture pre-dates the telescope, can also be seen in this painting. One is being held by the astronomer to the right of the one holding the astrolabe. The other is across the table from him.

Islamic astronomy

Turkey was the centre of the Ottoman Empire and this observatory, paid for by the sultan, was built for Taqi al-Din. The Islamic world has a long history of important astronomers and observatories, particularly during the thousand-year period dating from just after the founding of Islam in the early seventh century up to the European Renaissance. Taqi al-Din's observatory was built in 1577, when the Europeans were beginning to catch up, building their own observatories such as Tycho Brahe's in Denmark. The eighth-century 'House of Wisdom' in Baghdad was an important early centre for Islamic astronomy and learning. Here, great works were gathered from all over the world and translated into Arabic. The ideas found in the texts, including many from ancient Greece, were then brought together and developed. The lasting contributions made to astronomy during this period mainly concern mathematics, theories about how the heavens move, calculations and observations to support these theories and increasingly accurate measurements of position.

A by-product of this work was the allocation of names, including those for individual stars, many of which we still use today. If we look at some of the stars we have met so far we find Deneb, short for danab ab-dajaja or 'tail of the hen'. Though we see Deneb as part of the constellation Cygnus the swan, other cultures, including some pre-Islamic Arabs, saw it as a hen. As the Islamic scholars accumulated knowledge from different cultures they tried in many instances to bring these different traditions together. Interestingly, the Chinese tradition of astronomy, which remained separate from the Islamic and European traditions until relatively modern times, also depicted this constellation as a bird (at least they saw it as a magpie bridge, connecting lovers Niu Lang and Zhi Nu once a year in a story we'll revisit in Chapter 6).

Other stars that we've already met with Arabic names are Altair in Aquila, which comes from an-nasr aṭ-ṭā'ir, meaning 'the flying eagle'; Merak and Dubhe in Ursa Major, coming from *maraqq* and żahr ad-dubb al-akbar, meaning 'the loins of' and 'the back of' the Greater Bear respectively.

It is impossible to ignore the importance of Islamic astronomy in the history of the discipline. Until the seventeenth century many European astronomers were quite open about their debt to Islamic astronomers. Celestial globes showed the constellations named not only in Greek and Latin but also in Arabic, while astronomers occasionally named specific Islamic astronomers as their sources. In the seventeenth century, however, the ranks began to close as these European astronomers started to organize themselves as a profession.

Observatories

Regretably, the buildings that housed Taqi al-Din's observatories no longer exist, though there is a much earlier tower in the Galata district of Istanbul (since turned into a restaurant and café) which, it is claimed, was one of his observation posts. The observatory of his European contemporary, Tycho Brahe, was similarly destroyed within a generation, but there are many other astronomical tourist spots to seek out and visit, if you know where to look. On being asked to illustrate this book, Greg Smye-Rumsby, a planetarium lecturer at the Royal Observatory, immediately sent me his tour of London produced for his local amateur astronomical society. Overleaf is his map.

Here you can see the private observatories of several individuals – often visible as domes attached to otherwise quite ordinary, if large, London homes. You will also see several London landmarks, including the Monument and St Paul's. Surprisingly, these buildings

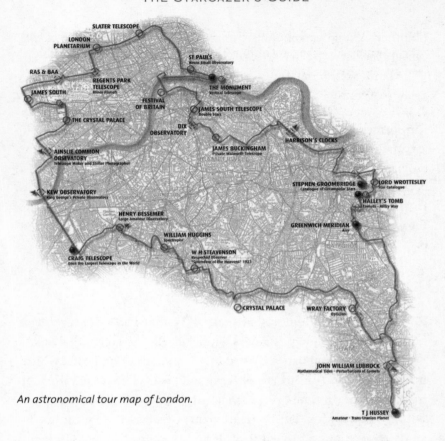

SLATER TELESCOPE

LONDON
PLANETARIUM

ST PAUL'S
Wren's Small Observatory

RAS & BAA

REGENTS PARK
TELESCOPE
Minor Planets

THE MONUMENT
Vertical Telescope

JAMES SOUTH

FESTIVAL
OF BRITAIN

JAMES SOUTH TELESCOPE
Double Stars

THE CRYSTAL PALACE

DIX
OBSERVATORY

AINSLIE COMMON
OBSERVATORY
Telescope Maker and Stellar Photographer

JAMES BUCKINGHAM
Private Walworth Telescope

HARRISON'S CLOCKS

STEPHEN GROOMBRIDGE
Catalogue of Circumpolar Stars

LORD WROTTESLEY
Star Catalogue

HALLEY'S TOMB
Comets – Milky Way

KEW OBSERVATORY
King George's Private Observatory

HENRY BESSEMER
Large Amateur Observatory

GREENWICH MERIDIAN
Airy

WILLIAM HUGGINS
Spectrohel

W H STEAVENSON
Respected Observer
'Splendour of the Heavens' 1923

CRAIG TELESCOPE
Once the Largest Telescope in the World

CRYSTAL PALACE

WRAY FACTORY
Optician

JOHN WILLIAM LUBBOCK
Mathematical Tides – Perturbations of Comets

An astronomical tour map of London.

T J HUSSEY
Amateur – Trans-Uranian Planet

have strong astronomical connections. The Monument was built by Sir Christopher Wren and Robert Hooke to commemorate the Great Fire of London of 1666, but it was also built as a telescope. Inside the column is a spiral staircase that winds around an empty cylinder. This cylinder forms the basis of the telescope. At the bottom is a laboratory which, when the Monument was being used as a telescope, could house a lens. This type of telescope is called a zenith sector, meaning it points only straight up – or to the zenith (another Arabic astronomical term). St Paul's, meanwhile, is the burial place of Sir Christopher Wren, an architect and astronomer.

Another of the landmarks mentioned on Greg's tour is the Craig telescope. Built by Vicar John Craig, a keen amateur astronomer from Leamington Spa, it once stood on Wandsworth Common. For permission to build Craig approached the 4th Earl of Spencer, Lord of the Manor of Wandsworth. He was persistent, as Spencer's lawyer, Charles Lee, remarked at the time to a friend:

> The Rev John Craig has been with me about the plot of ground for his 'monster telescope'. I hardly know what to do, he requires it within a week or his telescope will not be up in time to catch Mr Orion, who is now above the horizon.

By Orion he means, of course, the constellation Orion the hunter. Craig's perseverance paid off. With a tower 64 feet high and a telescope tube 85 foot in length his telescope attracted a good deal of public and press attention when it went up in 1852, though its heyday did not last long. Following questions about its accuracy and, after the deaths of Craig's wife and son, the telescope was abandoned and the whole site demolished in 1870.

A farther-flung site that I'm strangely fond of is in Slough. It commemorates the work of the Herschels, two of whom we've met already. First was William Herschel, whose discovery of the planet Uranus in 1781 was the first such discovery since antiquity. William's sister-turned-astronomer Caroline found several comets and became the first paid female astronomer. His son John became an important figure in nineteenth-century science and was one of the founding members of the Royal Astronomical Society. Their house and observatory no longer exist but the area around the site is filled with references to the family. There is a Herschel Street, a Herschel school, and, my personal favourite, a Herschel shopping centre.

Much farther afield there are many stupendous astronomical sites. Top of the list is probably Ulugh Beg's observatory in Samarkand, Uzbekistan. As a son of Shah Rukh, ruler of the Timurid Empire, Ulugh Beg was given charge of the city of Samarkand and decided to turn it into the empire's intellectual centre. As an astronomer and mathematician, his ambition was to build a huge observatory. Completed in 1428, it housed his sextant, made very large for maximum accuracy. Over the next nine years he used this instrument to compile a star catalogue, even now considered the greatest, most accurate and exhaustive of its age. It was superseded only just over a century later by Tycho Brahe's.

Through the constellations that help make up the Hercules story we've started to build up a greater picture of the night sky. The stars within them open doors into the golden age of Islamic astronomy and help to illustrate some of the problems faced by eighteenth-century European comet hunters. Now we turn to our own star, the Sun, which has been studied and looked at more closely than any other.

CHAPTER THREE

June and the Sun

WITH THE NORTHERN HEMISPHERE'S summer solstice and the shortest day fast approaching, June is a good time to do some daytime stargazing. During the day we can generally see only one star, the Sun, but a good look at it tells us all kinds of things about stars in general. For this reason, astronomers and stargazers alike have had a long fascination with this, our closest star.

> The Sun is a mass of incandescent gas
> A gigantic nuclear furnace
> Where hydrogen is built into helium
> At a temperature of millions of degrees
>
> The Sun is hot, the Sun is not
> A place where we could live
> But here on Earth there'd be no life
> Without the light it gives.

So said the band They Might Be Giants in their 1994 song, 'Why Does the Sun Shine', taking their words from a 1959 educational album. The ideas it expresses were not always held to be true, however. For example, the Sun was not always believed to be uninhabitable. Habitable Sun theories developed once European observers had started to see sunspots on the Sun's surface (these are the dark spots or patches you can see on the Sun if you look hard enough).

Viewing the Sun

You should never look directly at the Sun, but you can look at it through the appropriate filters. Terms to look out for when buying filters include 'mylar' or 'solarskreen'. Cardboard-framed glasses using these filters are often sold ahead of eclipses and can be bought through various websites. Make sure they have appropriate approval, a CE mark for example. They will make solar viewing safer, though it's still not a good idea to stare at the Sun for long. Look for just a few minutes at a time, give your eyes a rest, then look again.

Alternatively, you can project an image of the Sun on to a flat surface using a pin-hole. Simply make a pin-hole in a piece of card and hold it between the Sun and your flat surface, moving it around until you get a clear image. The smaller the hole, the shorter the distance you'll need between your card and the flat surface. An additional improvement you can make is to block out the light from anywhere but your pin-hole. One way of doing this is to position your pin-hole in the crack between closed curtains and let the image fall on a flat surface in your darkened room. (This is effectively the principle behind devices like the pin-hole camera and the camera obscura, both first described by Iraqi astronomer Abu Ali Al-Hasan Ibn al-Haitham in the eleventh century.)

Once you're safely viewing the Sun you can start to look for sunspots. We now know these dark patches are areas of relatively low temperature in the Sun's outer shell caused by high magnetic activity. The Sun, like the Earth, spins on its axis. And just like the Earth, the Sun has a north and south pole near each end of this axis. However, as a typical fully formed but still quite young star, the Sun is not a solid sphere like the Earth, but a sphere of gas, mostly hydrogen gas, held together by gravity and its magnetic field. Because the Sun is made up of gas, its shape, and in particular the shape of its outer layer called the corona, is not fixed but rather shaped and reshaped by the Sun's magnetic field. As the Sun spins, the gas around its middle moves round faster than that at the poles. This variation in the speed at which the gas travels creates little pockets of disrupted magnetic field, points at which the field strength is noticeably different from its surrounding area. The cause and effect relationship here is complicated, but the result is that where the magnetic field is very high, the temperature is very low, reducing from about 5800K (approx 5527°C) to around 4250K (approx 3977°C): this creates a sunspot.

The Maunders

In the mid-nineteenth century a German amateur astronomer called Samuel Heinrich Schwabe noticed that the number of sunspots fluctuated at roughly regular intervals. Every eleven years there is a maximum number of sunspots; this then declines, reaching a minimum about five or six years later. The numbers then go back up again, and so the cycle continues. The Swiss astronomer

JUNE

Rudolf Wolf followed this up, looking back in history at old records of solar observations. He discovered that the rule was true back to 1745 at least and probably (though records were fewer) to the early 1600s and Galileo. The problem was then pursued at the Royal Observatory, Greenwich, by the husband and wife team Edward and Annie Maunder.

Annie Scott Dill Russell met her husband, Edward Walter Maunder, when she started work at the Royal Observatory in 1891. Edward had been there since 1873 and was now head of the solar department. Annie was part of an experiment carried out by the Observatory in employing women. Four university-educated women were chosen to fill places in the lowest scientific post of 'computer' (the origin of the term computer). The post, generally held by fourteen-year-old boys appointed through examination, involved carrying out routine calculations on the observations made by the astronomers, who would turn them into something useful. In the end only three women were appointed (the fourth position was offered to Agnes Clerke, already one of the most successful astronomy writers of the late nineteenth century, but she turned it down).

Annie came third in her class at Girton College, Cambridge, twenty years after the college opened in 1869. At the Observatory, she was placed in the transit and solar departments. In the solar department she worked under Maunder on the daily photographs taken of the Sun from Greenwich, and sometimes from observatories elsewhere, including India and Mauritius. Though technically a computer she, like the Observatory's other lady computers, was allowed and encouraged to become involved in a wider range of duties. Annie gradually became an expert in the field of solar astronomy. At the end of her first year, Maunder put her name forward to become a Fellow of the Royal Astronomical

Society. At the time, the Society was all male (though Caroline Herschel had been made an honorary member while she was still alive); they refused her, and two other women, entry. She was eventually accepted in 1916 after returning to work at the Observatory during the First World War.

When she married Maunder in 1895 she left the Observatory but continued to work with her husband on solar observation. They travelled around the world viewing eclipses and studying sunspots. Annie made a catalogue of recurrent sunspots, and together they investigated historic sunspot cycles, discovering what has come to be known as the Maunder Minimum. Where Schwabe and Wolf found the rise and fall in sunspot numbers every eleven years or so (the figure fluctuates between nine and fourteen), the Maunders found, looking at the bigger picture, that the peaks also had their own cycle. They found the peaks to be very low in the period between 1645 and 1715, a time of intensely cold winters in Europe and North America. We now know this episode as the Maunder Minimum.

The reason for the eleven-year solar cycle was explained by George Ellery Hale, an American solar astronomer working at about the same time as the Maunders. He found that, like the Earth, the Sun can be seen as a magnet. The poles of this magnet swap round (north becomes south, south becomes north) every eleven years, then back again eleven years later, making a whole cycle of twenty-two years. This is called the Hale cycle and neatly links the rise and fall of sunspots with the Sun's changing magnetic field. As the Sun spins, the gas around the Sun's equator moves quicker than that at the poles; previously neat field lines (linking the poles in ever larger arches) get disrupted, looped, and produce sunspots. The longer this goes on, the more sunspots are produced until eventually the poles flip, and all that damage to the field lines starts to get undone.

The same magnetic activity that causes sunspots also causes other kinds of solar activity, including solar flares and solar winds. The latter cause something very dramatic to occur here on Earth, phenomena known as the Northern and Southern Lights, or the aurora borealis and aurora australis.

Northern and Southern Lights

The novelist Philip Pullman has brought the Northern Lights to popular attention recently with his series of modern children's classics, into which he weaves all kinds of historical ideas and scientific language. The alethiometer, for example, the instrument that helps guide the heroine, Lyra, to the North Pole, is a parody of a historical instrument called Crooke's radiometer. More significantly, the Arctic explorer who helps Lyra at the North Pole is called Scoresby, very probably after William Scoresby, a nineteenth-century Arctic explorer who travelled to the pole to investigate the Earth's magnetic field and so, in a roundabout way, the Northern Lights.

The Northern (and Southern) Lights – aurorae – are a fantastic though unpredictable phenomenon to observe. They can best be explained in terms of the interaction between the magnetic fields of the Sun and the Earth. They are best seen at night, close to the poles and close to sunspot maximum (when the number of sunspots visible at any one time on the Sun's surface is at its eleven-year peak). Because aurorae are unpredictable you are never guaranteed a view; all you can do is increase the likelihood of seeing them. The Earth's magnetic field is strongest at the poles; the Sun's magnetic field has its greatest affect on the Earth's at sunspot maximum. This means the closer you are geographically to the poles the more likely you are to see the aurorae at any time. Similarly, the closer you are chronologically to sunspot maximum the more likely you are to see the aurora even at some distance from the poles. At sunspot maximum (and the next is expected around 2011–12) it is sometimes possible to see the Northern Lights as far south as London.

The Earth can be though of as a magnet (this was first argued by Queen Elizabeth's physician, William Gilbert) with a north and south magnetic pole. (These magnetic poles are in a slightly different place from the poles we talked about in relation to the tilt of the Earth.) Like any magnet, they produce field lines joining one pole to another in ever wider arcs, and it is along these that the aurorae are sometimes seen arching. When charged particles from the Sun interact with the Earth's upper atmosphere and magnetic field, brightly coloured lights streak across the sky; these are what we call the aurora (named after the Roman goddess of the dawn).

Sunspot numbers are an indication of the state of the Sun's magnetic field. Their numbers increase as the magnetic field becomes more and more distorted. Besides sunspots, this distorted field can also result in solar winds, streams of charged particles sent out from the Sun's upper atmosphere. How these particles gain the energy needed to escape the Sun's gravitational pull is not well understood. What we do know is that at sunspot maximum the frequency of all the other phenomena associated with the Sun's distorted magnetic field also increases, including solar winds. This results in more aurorae and so a greater chance of seeing them for the casual observer.

The habitable Sun

Though early theories identified sunspots as cool spots, knowledge of the exact temperatures is relatively recent and rather rules out earlier theories on the habitability of the Sun. Western astronomers started to find sunspots in the seventeenth century, and for the next two hundred years or so a number of highly respected and influential astronomers suggested that there could be life just underneath the Sun's bright outer shell.

The French mathematician Jérôme Lalande suggested in his late-eighteenth-century book *L'Astronomie* that sunspots were evidence of mountain peaks below the bright outer atmosphere. William Herschel went further in his theories of a habitable Sun, arguing in 1801 that the Sun was 'a most magnificent habitable globe'. Conviction began to wane as the nineteenth century wore on, when astronomers began to look more closely at what the Sun was actually made of.

By looking at the Sun in this way it has gradually become clear that it is a type of star, not habitable in any shape or form. By a process called nuclear fusion, it creates heat and light as it turns hydrogen into helium at its core. While it's doing this it's called a main sequence star made up mainly of very hot gas, or to be more technically accurate, plasma. When the hydrogen at our Sun's core runs out, it will still produce heat and light energy as the helium is turned into other materials. At this stage, it is called a red giant. Once that phase ends it will throw off its outer layer, leaving a core surrounded by a glowing shell, a planetary nebula. The remaining core will later become a white dwarf. The size of the Sun and its relationship with the Earth will change dramatically as it becomes a red giant, burning away oceans and the atmosphere, but this isn't likely to happen for another four to five billion years. At the moment the Sun is still about 70 per cent hydrogen. Ultimately this means the Sun will never be habitable, now or at any time in its future.

The transit of Venus

The transit of Venus occurs when the planet Venus can be seen passing across the surface of the Sun during the course of several hours, as those who saw it in 2004 will be aware. Transits happen roughly twice every hundred years. The first to be observed was the 1639 transit, the second of the two to occur in the seventeenth century. There were two more in the eighteenth century, in 1761 and 1769, two more in the nineteenth, in 1874 and 1882, and none in the twentieth century. Our most recent transit was in 2004 and our next, which will be visible from anywhere with daylight for the full duration of the transit, will be in 2012.

In the late eighteenth-century the transiting of Venus prompted a number of expeditions around the world. Herschel's patron, George III, was an enthusiastic supporter and had his own observatory built at Kew from which to observe the phenomenon. The reason it interested seventeenth-, eighteenth- and nineteenth-century astronomers so much is because, theoretically, observations of the transit should make it possible to calculate accurately the size of the solar system. If you can measure from at least two points on Earth the exact time Venus first crosses the edge of the Sun, where on the Sun it appears to cross, and exactly when it leaves, you have all the information you need. Using trigonometry it should then be possible to work out the distance between the Earth, Venus and the Sun and so, since the ratios of the distances of every planet are already known, the size of the whole solar system. In fact this turned out not to be possible because of something called the 'black drop effect'. As Venus crosses the edge of the Sun, rather than looking like a neat circle, it appears elongated, like a drop. This means exact timings of when Venus crosses the edge are not possible. The reason expeditions continued into the nineteenth century was that astronomers were hoping to use the new technology, photography, to solve this problem. However, even in photographs, the circle turned into a drop.

The most famous of the eighteenth-century expeditions was James Cook's. Captain Cook is best remembered today as an explorer, 'discovering' new lands including the Hawaiian Islands and the east coast of Australia, and as the first European to circumnavigate New Zealand. However, none of his expeditions to the Pacific would have taken place without the transit of Venus, which was the official reason for his first voyage (though he was given secret instructions to continue on after the transit and explore new lands for the British). For the 1769 transit of Venus, Cook was sent to Tahiti. Astronomers were sent to Hudson Bay in Canada and Baja California in Mexico, and the transit was also observed from St Petersburg by the Czech astronomer Christian Mayer at the invitation of Catherine the Great. Nothing conclusive was drawn from these expeditions, not least because the variety of apparatus used made the observations difficult to compare. Lessons were, however, learned for the future.

The nineteenth-century official expeditions determined to leave nothing to chance. George Biddell Airy, Astronomer Royal at the Royal Observatory, was placed in charge. For his five planned expeditions he ordered five sets of identical apparatus, including photographic equipment, to make sure that this time the observations would be accurately comparable. He even had them all set up in Greenwich Park ahead of time just to check everything was in working order.

Groups of astronomers and members of the Admiralty were then shipped off to Egypt, New Zealand, Honolulu, Rodriguez Island off the coast of Madagascar and Kerguelen Island in the Indian Ocean. In each case they were sent with detailed instructions on how to behave in their designated location. In New Zealand they had to persuade locals not to burn grass (part of the annual work on local farms) until after the transit had taken place so that the smoke wouldn't obscure their view. In Honolulu they narrowly missed losing all their apparatus when a storm blew down a coconut tree that just missed their camp.

This photograph, taken by Edward Maunder, is one of my favourites. The slightly bored-looking astronomers sit in a box, as though they're about to be sealed up and shipped abroad along with their equipment.

Eventually, observations were taken and carried back to Greenwich. The astronomers had been trained to use a new photographic technique invented by the French astronomer Jules Janssen. The initial purpose of this precursor to cinematography was to ensure that the exact moment Venus appeared to meet the edge of the Sun was not missed. Sometimes referred to as the photographic revolver, it allowed many pictures to be taken in quick succession around the circumference of a circular photographic plate. It was thought that the black drop effect was the eye getting confused and so could be solved with photography. This proved not to be the case, and the circle of Venus still appeared as a drop. Why the black drop effect occurs is, even today, a mystery.

These historical transits were very much the domain of the professional and serious amateur astronomer. The 2004 transit was something of a phenomenon in its own right, the first to be regarded as a popular tourist attraction.

I spent the 2004 transit at the Observatory along with literally thousands of visitors – all looking up at the Sun through eclipse glasses or telescopes with filters on them or looking down on to projections of the Sun's surface. The transit took a typical six hours, beginning at 5.20 in the morning. The museum as a whole (the Royal Observatory is part of the National Maritime Museum) took some persuading that this was a spectator sport and that people would come to see it at that time in the morning. We were right: people started queueing even before the gates had opened. Elsewhere the transit was viewed by large numbers of people all over the world, often with the help of amateur astronomy societies and professional observatories. Filmed by TV crews, it even made the front page of a number of national newspapers. The next transit is on 6 June 2012 and will be visible in its entirety from

north-western North America, Hawaii, the western Pacific, northern Asia, Japan, Korea, eastern China, the Philippines, eastern
Australia and New Zealand. Given the popularity of the 2004
event, it's likely to be quite a tourist attraction and well worth the
excuse to travel.

Eclipses

Eclipse expeditions as tourist attractions have a much longer history. Astronomers began making journeys to view eclipses in the
mid-nineteenth century, and eclipses still offer something abso-
lutely spectacular to see. In a total solar eclipse the Moon appears
to block out all but the very edge of the Sun. The sky goes dark,
birds fly furiously across the sky thinking it's night. There is an
eerie silence as the local fauna try to work out what's going on. If
you're lucky, this might last a few minutes. You then get what's
called Baily's beads, a phenomenon named after the nineteenth-
century banker and amateur astronomer who first noted their
appearance. Baily's beads are the little beads of light that appear on
one side of the disk as the Moon starts to move away from its position obscuring the Sun and light from the Sun starts to shine
between the Moon's craters. Slowly, more and more of the Sun
becomes visible until it's all over, at least for another few years.

Total solar eclipses are visible from somewhere on Earth
roughly once every eighteen months. The Earth orbits the Sun in
one plane, the ecliptic. The Moon orbits the Earth in another, tilted
just slightly in relation to the ecliptic. This means that the Sun and
Moon line up only at the two points (called nodes) when, during
the Moon's 27.3-day orbit, the Moon crosses the ecliptic. If it's a
new Moon when these paths cross there is a solar eclipse. (A new

A total solar eclipse photographed in Dundlod, India, by Fred Espenak on 24 October 1995

Moon is when there's no Moon visible at night; it's the opposite of a full Moon.) Because both the Earth and the Moon's orbits are elliptical, their distance from the Earth varies, and this alters their apparent size. For a total eclipse, the Sun and Moon need to appear from Earth to be the same size (this calls for the Earth to be near its farthest point from the Sun, and the Moon its closest point to the Earth). If the sizes appear different, the Moon does not completely obscure the Sun; this is called an annular eclipse. Though the eclipse is still impressive, you don't have some of the features we tend to associate with total solar eclipses, such as the visible corona (the Sun's outermost layer) or Baily's beads.

Lunar eclipses occur when there is a full Moon at the moment the Moon's path crosses the ecliptic. Gradually, over a period of several hours, you see the Moon disappear, as the Earth comes between it and the Sun, stopping the Moon reflecting the Sun's light. When the three bodies – the Moon, the Earth and the Sun – are completely in line, the Moon appears red. This is because the light from the Sun now has to pass through the Earth's atmosphere to get to the Moon and all but the red light is scattered away. This scattering of light by our atmosphere is also the reason the sky appears red sometimes at sunset. Unlike solar eclipses, you don't have to be within a narrow path to see a lunar eclipse, you only need for it to be night and for the Moon to be above the horizon at the time it happens. This means there are more opportunities to see them. Typically, a lunar eclipse occurs roughly once every six months, though these are not always total and not always visible for the same half of the world (since seeing it depends on the Moon being visible from your location at the time).

Total solar eclipses occur in the same place only once in several generations. They are therefore often imbued with special powers. In one ancient Chinese legend, the Sun had been swallowed by an invisible dragon. The solution was to make lots of noise to frighten away the dragon. Eclipses were also regarded as omens by both the ancient Chinese and Babylonians. A coming eclipse was supposed to indicate an important event in the life of the ruler, and so it became important in both cultures to work out how to predict them.

A Hindu tradition, still practised in some areas of India today, much to the surprise of modern European eclipse chasers, is that pregnant women must be protected from viewing an eclipse to prevent the child being born with some disability. As I write this book, I am five months pregnant. I have been strongly advised not to drink even a single glass of alcohol, or to eat pink meat, soft cheese or peanuts. All these consumables carry a very low, sometimes entirely unproven risk to the baby, but I avoid them because I fear my luck might turn if I slip. No one has advised me to avoid eclipses, but I can fully sympathize with these Hindu women and their ancient superstition.

Eclipses became popular as a spectator event once astronomers began to pass on predictions to the general population. Edmund Halley was an early promoter, publishing a map showing the path an eclipse in the 1720s was expected to take over the British Isles.

Halley's motives for encouraging popular interest in eclipses were not entirely disinterested. His aim was to improve the accuracy of eclipse prediction by trying to amass as many observations as possible. What he found was that the path for the eclipse was much narrower than he had predicted, and so it was visible from fewer parts of the country.

Halley's promotion of eclipses popularized astronomy. In this he was helped by the recent death of Newton, which led to a growing interest in astronomy as people began to feel they ought to understand better the work of their national hero. However, it was not until long-distance travel became a viable option for the general public that eclipse tourism really took off. By the late nineteenth century, and certainly by the twentieth century with the advent of large railway networks and cruise ships, it had become possible for people to travel to an eclipse in significant numbers. Eclipse tourism became a phenomenon.

At first large groups of publicly funded astronomers started to travel. They travelled around Europe, to Turin in 1842 and Spain in 1860 where the first photographs of an eclipse were taken by businessman, amateur astronomer and photographer Warren de la Rue. There were also less formal outings towards the end of the century. In America the famous astronomer Maria Mitchell took some of her students, all female, from Vassar College on an expedition to Burlington, Iowa in 1869 and another group to Denver, Colorado in 1878. They travelled by train and pitched their tents at their final destination ready to view the eclipse and write up their observations. Not only did this give her students a taste of independent travel, it was also pioneering in educational terms. Field work was hardly part of any students' education at the time, male or female.

For my own part, I'm ashamed to say I've travelled to see only one eclipse. I'm more of an armchair eclipse chaser. I like the idea, and I like reading accounts and looking at pictures of the intrepid travellers of the past, but I'm very bad at actually doing anything about it myself. My trip was advertised in the back of an astronomy magazine and, unlike the eclipse cruises, cost only £50. It was to see the 1999 eclipse from France. Though this same eclipse was visible from England I reasoned that France might offer a better likelihood of clear skies. We were an amiable group, everyone was a little over-excited. We spent an awfully long time on the coach, travelling until late in the evening and then stopping at a hotel for the night before more travelling the next day. But then we were there – an empty field. Hundreds of other coach parties and individuals piled in. My boyfriend and I opened the bottle of cheap red wine we'd bought in Carrefour on one of our brief stops, poured it into our plastic wine glasses and sat down, our eclipse glasses at the ready. The day was quite cloudy but nevertheless the skies darkened, the birds flew furiously across the sky and silence fell. And the clouds did part briefly for us to view totality, which was every bit as spectacular as we'd hoped, probably enhanced by the festival-like atmosphere of having so many people in a muddy field. And that was my first and, to date, last eclipse expedition. Perhaps 2009 will be different.

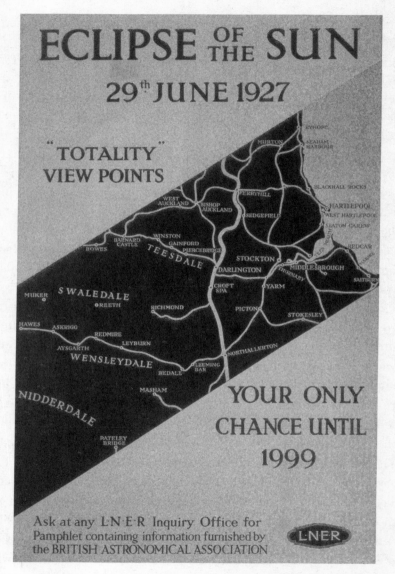

A 1920s LNER poster advertising train travel as a means of getting to the eclipse site

On 22 July 2009 there will be a total solar eclipse visible from India, Nepal, Bangladesh, Bhutan, the very edge of Myanmar (Burma), China (including Tibet), and various islands in the western Pacific. Detailed maps for those considering a trip can be found on the NASA website. You could always combine it with some of the other astronomical sights in the region, such as the observatories of Jai Singh in Jaipur and Delhi in India, the Beijing ancient observatory in China or the Maha Wizaya Pagoda in Yangon in Myanmar. In this temple, built in 1986, lights representing the stars come on in each constellation as you press a switch. My very old friend (we were at playgroup together), who now lives in Burma as a Buddhist nun, brought this one to my attention. I haven't been able to find out much more about it, though it sounds as if it is well worth a look. The 2009 eclipse could form the basis of quite an astronomical tour, bringing together centuries of fascination in our most easily identified and visible star, the Sun, and, like all solar eclipses, leading us into a world of surmise, wonder and awe, an eerie celebration of still incompletely understood powers beyond ourselves.

JUNE

CHAPTER FOUR

※

July and Bayer's Menagerie

SINCE THE HEIGHT OF SUMMER brings little in the way of dark skies for the northern hemisphere, July is a good time to concentrate on the constellations of the southern hemisphere. Star maps from the ancient Greek and Islamic worlds show very few constellations in the southern hemisphere: they couldn't see that far. The people then living in Southern Africa, Australia, Southern America and other inhabited lands south of the equator had their own constellations, but few if any star maps survived, if indeed they were made at all. Some work has been done in recent years to try to rediscover and publicize some of these lost constellations, but the official constellations are now fixed.

Australian constellations

In the Elvina Track engraving site in Ku-ring-gai Chase National Park, near Sydney, Australia, there is an engraving of an emu (see page 84). This represents a constellation covering roughly the same area of the sky as the official constellations Crux to Scorpius. It is a constellation found in many aboriginal cultures and marked out in the night sky not by the stars, as all the constellations we've met so far have been, but by the dark and hazy clouds between the stars. The Guringai people who created this engraving were wiped out after the arrival of the British, and so its significance was lost. But recently it has been rediscovered that the emu engraving lines up with the sky emu at the exact time of year in which the emu lays her eggs, and its true astronomical significance was recognized.

The emu constellation was not unique to the Guringai people. It is found in a number of Australian Aboriginal cultures and languages groups, some of which still exist today. It is worth noting that there are between three and four hundred Australian Aboriginal language groups, all with their own language, culture, stories and songs. Some of these stories overlap but none is common to all groups. Experts in the fields of anthropology and archaeoastronomy have recently been talking with surviving Aboriginal cultural groups about their culture and the place of astronomy and constellations within it. More detail on this can be found at the Australian Aboriginal Astronomy website.

To find the emu in the night sky you should look first for the Southern Cross (on the star chart if you aren't lucky enough to be Down Under). Next to this you should find a dark patch of sky in the otherwise very bright band known in the West as the Milky Way. The body of the emu stretches out along the Milky Way, taking in the Coalsack nebula, until its feet meet the stars known to us as the constellation Scorpius. The photograph overleaf, coupled with the star chart, should help. We will meet the Milky Way and Scorpius in more detail in later chapters.

This photograph, taken by Barnaby Norris, the son of the Aboriginal Astronomy Project leader Ray Norris, beautifully shows the engraving of the emu below the night sky's emu.

Not all the constellations created by Australian Aboriginal cultures use the space between the stars rather than the stars themselves. Some, such as the Mallee-Fowl and Julpan the Canoe are more conventional. The Mallee-Fowl constellation roughly corresponds with our constellation Lyra. As this constellation (visible in the July sky) starts to disappear from the night sky around October this is a signal to the Boorong people to gather the eggs of a particular bird found in the Mallee region of Australia. Julpan the Canoe is a constellation of the Yolngu people from Arnhem Land, an Aboriginal reserve in the Northern Territory of Australia. The stars making up this constellation are roughly the same as those in our constellation Orion, with the three stars of Orion's belt marking out the central width of the canoe. The story tells of two brothers out on a fishing trip. The brothers caught and ate a particular fish forbidden in Yolngu law and, as punishment, the Sun sent a jet of water that pushed the brothers and their canoe up into the sky.

Australian Aboriginals were traditionally hunter-gatherers living and travelling in groups. This is clearly reflected in the types of stories they created about the sky, which describe seasonal changes and provide a calendar of natural events to help the storytellers and listeners find plentiful seasonal food. They also give strong messages about appropriate behaviour to aid the smooth running of the group as a whole.

It is interesting to note that many aboriginal cultures include star colours in their stories – that is, they differentiate between stars based on an observed difference in colour. Colour is regarded by modern astronomers as one of the clues telling us what a star is – the stage of its lifecycle, how hot it is and how it moves in relation to us and to the stars near it. For us, with our brightly lit skies and lack of practice in naked-eye stargazing, the idea that people could once see these subtle differences without the aid of a telescope is amazing. In Taurus, for example (the zodiac is visible from both northern and southern hemispheres), certain Aboriginal groups regard a cluster of stars astronomers call the Hyades as two rows of girls. There's a row of red stars and another of white stars, the red being the daughters of the red star Aldebaran, also in Taurus. We now know Aldebaran as a red or at least orangey giant star with a redder tinge to it compared with other stars; this can just about be made out with the naked eye. Aboriginals were well ahead of us in observing this difference.

Many of these constellations and their associations had existed for tens of thousands of years before Western anthropologists started to write them down. They survived some 40,000 years through word of mouth and a continuation of a culture that still understands and still has a place for the associated stories and messages.

South African constellations

In South Africa there also exist alternative constellations to those we use today, created by the various tribal groups living there before the arrival of Europeans. The Southern Cross was seen by many different tribes as marking out the heads of a group of giraffes. As their heads skim the tops of the trees in October it signals the time to end planting. The star Canopus in the southern constellation Carina is another with significance for a number of different tribal groups. The Sothos people believed that whoever spotted the star first would have good luck and riches for the rest of the year. The next day the year's luck for the whole tribe would be predicted using divining bones. The first of the Venda people to see the star in the morning sky climbed to the highest point and blew a horn. The Mapeli would also signal their first sighting by making as much noise as possible, loud enough to be heard in nearby villages.

In 1998 a project was established as part of South Africa's first Science and Technology Year to use some of these rediscovered constellations to promote science throughout South Africa. A poster was produced to illustrate some of the constellations used by native South African tribes. As in Australia, we have recently seen significant changes in how the native populations of South Africa are viewed. There is an attempt to learn more about these cultures and incorporate this knowledge into the cultural identity of the country as a whole. Astronomy and stargazing, as ancient cultural barometers, have an important part to play in this process.

A good many of the constellation stories collected and disseminated by Westerners concern constellations known within the northern hemisphere. The truly southern hemisphere constellations, those around the south celestial pole, seem to have been much less well documented. Whether this is a reflection of the interests of the astronomers and anthropologists collecting the information or of the distribution of stories across the night sky within native cultures, I cannot say.

South America

While Australia's Aborigines have remained hunter-gatherers and South African tribes have consisted mainly of settled farmers, the Mayan, Aztec and Incas of South America created larger and consequently more structured civilizations. The Mayans had a courtly class that included astronomers. Their task was to watch the heavens to predict the future and to find omens. They made and recorded very accurate observations and used mathematics to extrapolate forward from these observations. They had a zodiac, similar though not the same as ours, made up of animal constellations through which the Sun, Moon and planets appear to travel. They could predict eclipses and, crucially for the preservation of their lasting legacy, they wrote things down. Because the sixteenth-century Spanish settlers destroyed most of what these ancient civilizations had written, here too a process of rediscovery is currently being undertaken. For lasting records, it is really the Incas we need to turn to.

Between 1438 and 1533 the Inca Empire centred on an area of the Andes now divided between Peru, Ecuador, Bolivia, Argentina, Chile and Colombia. We have already met the Inca ruin, Machu Picchu. Among many other things, the ruin demonstrates that, like the Mayans and Aztecs, the Incas were interested in those astronomical calendrical markers, the solstices and equinoxes.

The surviving Quechua-speaking people of South America still use some constellations that can be traced back to the Incas. Like the Australian Aboriginals, the Quechua-speakers include in their constellations dark clouds in the Milky Way as well as the more familiar dot-to-dot star constellations. Among these constellations are various animals – the fox, the partridge, the toad and the great anaconda – each of which first rises in the night sky when that animal is most active on Earth. For example, Yutu, the constellation of the partridge, appears in the sky from early September until mid-April; farmers use this as a warning to protect their crops against the bird in its most active period. Similarly, the appearance and disappearance of the toad constellation indicates the coming and going of the rainy season.

Not all South America was populated by complex and bureaucratic civilizations, however. The Desana or Tukano people from the Amazon were originally hunter-gatherers. The Desana use their understanding of the night sky as a means of structuring their collective lives. They have a hexagonal constellation that marks the equinoxes. The hexagon is made up of the stars Procyon (in Canes Minor), Pollux (in Gemini), Capella (in Auriga), Canopus (in Carina), Achernar (in Eridanus) and another fainter star in Eridanus. When this hexagon sits above the Earth at sunrise or sunset it marks the equinox. This hexagon shape features strongly in the way Desana society is organized, and this is said to be a reflection of what is seen in the sky rather than the other way round. So, the longhouses in which the Desana live are hexagonal; men live in different areas within these longhouses depending on where they are within the six stages of life, from naming (Capella) to initiation (Pollux) to marriage, and so on; while women similarly move through a six-stage cycle until marriage, then they follow their husband.

For the most part the constellations originating in Australia, Africa and South America have been ignored in any official descriptions of the sky. The constellations that we use today were on the whole created by European astronomers and explorers from the seventeenth and eighteenth centuries.

Bayer's menagerie

Johann Bayer was an astronomer and lawyer from Bavaria, in modern-day Germany. In 1603 he published his famous star atlas, Uranometria, the first to include both the new southern constellations and the well-established Ptolemaic northern hemi-sphere ones. The name was taken from the Greek — Urania was muse of the heavens; Uranometria translates as 'measuring the sky' or the heavens. Bayer did not invent the new constellations himself or travel to the southern hemisphere to view them but rather took his information from a catalogue produced by the Dutch navigator Pieter Dirkszoon Keyser. Keyser had in turn taken his information from the work of two Italians, Amerigo Vespucci and Andrea Cor-sali, and the Spanish cosmographer Pedro de Medina. Vespucci was the Italian merchant and explorer best know for giving his name (Amerigo) to the Americas. Florentine Corsali, famed as the first European to describe the Southern Cross in print, was an employee of the Medici family and, for reasons that are unclear, a traveller on a Portuguese boat. De Medina was the Royal Cosmographer to the Spanish court. Between them they created a set of constellations for the southern hemisphere that were truly European.

Vespucci, Corsali, de Medina, Keyser and Bayer all lived through the European age of exploration. Vespucci and Corsali were both roughly contemporary with Christopher Columbus, and, like him, were sent out as explorers to discover lands (and resources) previously unknown to Europeans. As they ventured away from familiar lands they found themselves under unfamiliar stars and duly made a note of what they saw. De Medina, working a little later in the mid-sixteenth century, was primarily a navigator. For him, knowing accurately the position of these newly discovered southern hemisphere stars was essential for helping to establish the ship's position at sea and the direction in which they were travelling. The job of mapping these new stars, however, was still seen as of only minor importance compared with the more readily heroic work of discovering new lands. Keyser's experience was different.

The Low Countries (roughly speaking Holland and Belgium), where Keyser hailed from, already had a strong tradition of very accurate map- and globe-making. Keyser was sent as navigator on one of the late-sixteenth-century expeditions that helped to establish the Dutch East India Company. Before the voyage he was trained by one of the Company's founder members, the astronomer and cartographer Petrus Plancius, to map the southern stars, and on his return delivered his observations to his teacher. Plancius turned them into twelve constellations which he had published on a globe in 1597. Six years later Bayer used these constellations, together with the very latest celestial cartography of the northern hemisphere, in his map.

These new constellations, often described for brevity as Bayer's constellations, consist of Apus (the bird of paradise), Chamaeleon (the chameleon), Dorado (the goldfish), Grus (the crane), Hydrus (the lesser water snake), Indus (the Indian), Musca (the fly), Pavo (the peacock), Phoenix (the firebird), Triangulum Australe (southern triangle), Tucana (the tucan) and Volans (the flying fish). Most of these you can see on the illustration from Bevis's star atlas.

On the Bevis illustration, shown overleaf, you will see the Southern Triangle just above the Southern Cross, farther along the Milky Way. To the left of that is Apus, then Pavo. Tucked in behind Pavo is Indus, apparently pointing an arrow at Pavo. Travelling further round the picture in an anticlockwise direction you will meet Grus, Tucana and Hydrus. Below these is the Phoenix rising from the flames. Finally, at the bottom centre of the page, is Dorado and, in front of him, Volans, complete with wings. The strange rat-like creature between Volans and Apus represents Chamaeleon, while Musca is the tiny constellation, just in the Milky Way between the chameleon's nose and the Southern Cross.

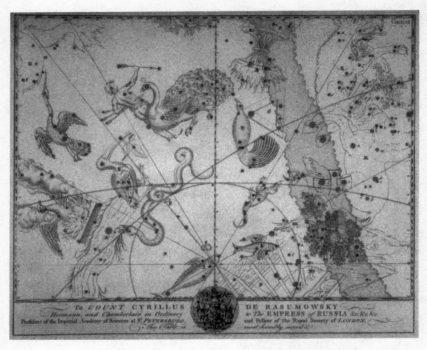

Various southern hemisphere constellations from Bevis's atlas.

All these constellations are relatively close together in the night sky. Once you've found the Southern Cross and the Milky Way you're well on your way. Some of them are made up of quite faint stars: the Chamaeleon in particular can be very difficult to make out. Apus and Volans aren't much better, since their brightest stars only approach apparent magnitudes of 4. Grus, Pavo and Triangulum Australe are clearer, each with at least one star of apparent magnitude nearing 2, and are consequently probably the best to begin with. They all run roughly in a line. First, find Triangulum Australe on the Milky Way, just a little way from the

Southern Cross. You can then use this to find Pavo and then Grus. Indus can then gradually be made out, with practice and careful study of the modern star charts found in this book. (You might also like to use Google Earth, which gives you a good overview of where everything is in the night sky and allows you to move around and search for particular constellations. If you zoom out from one of these Bayer constellations you can see where the others sit in relation to it without Bevis's classical illustrations to distract you.)

Once you've found Triangulum Australe, Pavo and Grus you can then start to look for the next line of constellations: Apus, Tucana and Phoenix. Be aware that these three, and Apus in particular, are a little fainter than those you started with.

You will notice that the Chamealeon does not look very much like a chameleon while the fish with wings and the bird with no legs just look odd. Naturally, Bevis hadn't seen any of these animals and had to work from descriptions, or a sketch if he was lucky, and from his imagination. (Probably the most famous example of this kind of misinterpretation of exotic animals by artists relying on explorers' descriptions is Albrecht Dürer's 1515 woodcut of a rhinoceros; so Bevis is at least in good company.)

Some of Bevis's misunderstandings about the animals in the sky came from the descriptions from explorers. Apus, for example, was based on a bird native to Papua, New Guinea. When these birds were sold at market to Europeans the Papua New Guinean traders would cut off the birds' legs, leading the Europeans to believe that the birds never had them to begin with.

The curious and exotic animals these explorers chose as their constellations reflect a broader interest in the exotic found all over Europe at the time. King Manuel I of Portugal had a rhinoceros and elephants brought over from India. Other royal courts also had menageries made up of exotic creatures and cabinets full of curious objects brought over from these newly discovered lands.

Gradually, the royal interest in collecting exotica filtered down. Private collections of rare and curious artefacts, including stuffed birds and beasts, became fashionable commodities in seventeenth-century Europe. Some of these larger collections would become the foundations for many of our museums. Individual collectors included people like Hans Sloane in London, who made at least some of his fortune in chocolate and whose collection became the foundation for the British Museum.

Interest was not confined to the collectors. Collections like Sloane's attracted large numbers of curious visitors. To the salons, the literary and philosophical discussion groups often run by aristocratic ladies in and around Paris, exotic creatures were sometimes introduced. They were often sent directly to the salonnière (the lady running the salon) by a friendly explorer, as a talking-point.

Salons evolved in the seventeenth century as social gatherings among the aristocracy and new rich. The idea behind them was to encourage intellectual discussion and expose the men attending them to the civilizing influence of women, thus making the ruling classes as a whole more refined. Science was popular for being very visual. Objects of curiosity could be brought for discussion from around the world and experiments could be performed, providing the guests with a sometimes dramatic display.

It was in the spirit of this widespread fascination with creatures and indeed people (the constellation Indus, after all, is supposed to represent a Native American) from newly discovered lands that Bayer and the explorers named these new star groupings.

By now it should be clear that constellations are made up: there is no scientifically right way of organizing the star patterns we see in the sky. Different cultures have grouped them in different ways and created different stories to help remember them and what they signify. This arbitrariness was not lost on Bayer and his contemporaries.

The unofficial sky

While Bayer was marking out the new constellations on his star atlas, Uranographia, his friend and fellow lawyer, Julius Schiller, was busy devising an entirely new system for the whole sky. Schiller's great project, which he completed the year he died, in 1627, was to produce a star atlas with characters from the Bible to replace all the traditional constellations: in other words, he set about Christianizing the sky. The twelve zodiac constellations became the twelve apostles; the northern hemisphere constellations were replaced with characters from the New Testament; while the southern hemisphere became populated with characters from the Old Testament. The system was never widely adopted, though something like it was also produced by teacher and globe-maker Erhard Weigel, at the very end of the seventeenth century.

Weigel's globe, now part of the collection I used to curate at the National Maritime Museum, is a metal globe with the constellations in relief. Where Schiller used Christian figures to replace the more traditional constellations, Weigel used heraldic symbols. Ursa Major, for example, is represented by the heraldic symbol for Latvia, as designed by Weigel. The purpose of this globe was to teach people about the constellations. This idea of renaming constellations is an educational concept that has been used ever since;

it is still a popular exercise with school groups to get the children to make and draw their own constellations based on the stars they can see.

On similar lines, the Greaves and Thomas globe replaces all the modern, 'official' constellations with characters from *Alice in Wonderland*, Jabberwocky and *Alice Through the Looking Glass*. The characters are chosen to complement in some way the original constellations they replace, so, for example, Alice replaces Virgo the virgin, Gemini (the twins), is replaced by Tweedledum and Tweedledee. Bayer's constellations, as part of the officially recognized eighty-eight constellations of the night sky, are also included. The Chamaeleon is replaced by the Tove from Jabberwocky, a strange creature that changes its shape from badger to lizard to corkscrew. Tucana is also replaced by a character from Jabberwocky, becoming the equally large-beaked Borogove bird. Musca becomes the bread-and-butter fly from *Through the Looking Glass* while the Phoenix becomes the Griffin from *Alice in Wonderland*. Like Weigel's, the purpose of this globe is to educate. It was not designed to suggest that these new constellations should be officially adopted by astron-omers around the world. Neither was it intended to comment on the Alice books by suggesting any deliberate analogy between Lewis Carroll's characters and the constellations. Its purpose is to make us think about the night sky in a different way, to stop us taking for granted the official constellations.

New planets and Magellanic Clouds

A number of stars other than our Sun have been found to have planets orbiting them. The current estimate stands somewhere between two and three hundred stars. Most of these, unfortunately

In this globe, made by Greaves and Thomas, the traditional constellations have been replaced by characters from Lewis Carroll's Alice in Wonderland.

for the stargazer, are very faint and are invisible to the naked eye. (Not of course that we would ever be able to see the planets directly – not even the most powerful telescopes can do that.) In the constellations we have met in this chapter, there are no fewer than sixteen stars known to have at least one planet orbiting them, that is, sixteen other solar systems. Grus and Phoenix each have three stars, each with its own planet, all of which are too faint for the human eye to see (with the possible exception of **T**-Gruis, but even this has an apparent magnitude of 6, so your eyesight would have to be fantastic).

Better bets for stargazers are a couple of relatively close galaxies, the Small and Large Magellanic Clouds found in Tucana and Dorado respectively. These appear to the naked eye to be small bits of the Milky Way that have broken off and drifted away. They're very bright and so consequently relatively easy to see with the naked eye – the Small Magellanic Cloud has an apparent magnitude of about 2.3, the Large Magellanic Cloud of about 0.1.

A galaxy is a very large group of stars (millions, billions or even trillions), dust, gas and dark matter, held together by gravity, that all rotate around a common centre. They typically contain all the different stars and groups of stars we've already met – nebulae, main sequence stars, red giants, white dwarfs, double stars and star clusters – and they come in a range of shapes and sizes. The Milky Way is our galaxy and most of the star-like objects we can see in the night sky are part of our galaxy. A few however, such as the Magellanic Clouds, are neighbouring galaxies. At 210,000 and 179,000 light years away respectively, they are almost our closest neighbouring galaxies. Only a dwarf galaxy in the constellation Sagittarius is closer. Together with many more galaxies, including some that are relatively well known thanks to their popularity in science fiction – the Andromeda Galaxy, the Triangulum Galaxy and our own Milky Way – they form our local group of galaxies. The Sagittarius, Andromeda and Triangulum galaxies are all named after their constellations. The Small and Large Magellanic Clouds, on the other hand, are named after the Portuguese explorer credited with their discovery.

Fernando de Magellan learned astronomy at the Portuguese court and began his working life by taking part in an expedition run by Francisco d'Almeida. He served in Portuguese India in 1511 (from where a rhino would soon be sent to the Portuguese King) and in Morocco in 1513. He then fell out of favour with the court and went to Spain, where he found a wife and a new patron, Charles I of Spain. Charles provided him with the ships and men needed for his expedition, and off he went in 1519, becoming the first to circumnavigate the globe. Although he died on the trip, as did most of the men travelling with him (only eighteen survived out of 265), the voyage was regarded as a success.

Leaving aside for a moment the fact that many people and indeed whole cultures living south of the equator knew about these 'clouds' long before Magellan discovered them, he was still not the first to have done so. Al-Sufi was probably the first to mention the clouds in print, referring around 964 CE to knowledge of the sky held by people living in the southern part of Arabia. Al-Sufi was a tenth-century court astronomer based in Isfahan in modern-day Iran. He was a very important astronomer and mathematician whose best-known work is probably his very accurate star catalogue. In that catalogue, the Book of Fixed Stars, he calls the Large Magellanic Cloud Al Bakr, 'the White Ox', pointing out that it was visible from the Strait of Babd al Mandab, not from Northern Arabia or Baghdad. (The Strait of Babd al Mandab is the strip of water around the coast of Yemen, separating Asia and Africa and linking the Red Sea with the Indian Ocean.)

However, neither Al-Sufi nor Magellan knew they were looking at galaxies. It is only very recently that galaxies beyond ours have been shown to exist, and for this we have to thank American astronomer Edwin Hubble. Hubble's name is now associated with the

Hubble Space Telescope which orbits the Earth outside its atmosphere so that it can observe the sky without the atmosphere getting in the way and blurring the images. You can see the telescope from Earth if you know when and where to look. It looks like just another star, except that you can see it moving across the night sky. Look up the timings for where you are at www.heavens-above. com, and then it's just a question of looking. (Quick tip: search the site's database for HST rather than its name in full.)

Edwin Hubble worked under George Ellery Hale (whom we met in Chapter 3) at the Mount Wilson Observatory in California. He proved in 1925 that some of the objects thought of as nebulae within our own Milky Way were in fact galaxies. After that groundbreaking and career-making moment, he went on to create the classification system for galaxies that we still use today. This system is based on how galaxies look – like spirals, barred spirals (spirals with a bar in the middle making a sort of 'S' shape), ellipticals or lentils (called lenticular galaxies). Because the system is based on how they appear there is an element of judgement and subjectivity about the classifications. The best way to overcome this is for lots and lots of people to look at each galaxy and classify it, then use the classification most people agree on. Not only is this important for astronomy, but nowadays it also provides the star-gazer with an indoor rainy day activity. Galaxy Zoo, www. galaxyzoo.org, is a website devoted to just this problem. Stargazers are invited to assign classifications to a series of galaxies shown in pictures.

Not all galaxies have to go through this process in order to be classified, but only the more recent discoveries. The Milky Way and the Magellanic Clouds do not have to be subjected to this process; their classification is well established. The Milky Way is a barred spiral galaxy (with our Sun situated along one of its arms). The

Small Magellanic Cloud is a dwarf galaxy that was once a barred spiral but is now an irregular shape thanks to some distortion caused by its close proximity to the Milky Way. The Large Magellanic Cloud is also irregular.

The galaxies and constellations of the southern sky, many of which reflect the interest in the exotic and in collecting and classifying which swept across Europe in the seventeenth century, bring us to another new set of constellations, Lacaille's, added over a century later.

CHAPTER FIVE

August and Lacaille's Mountain

ONCE YOU START TO LOOK, stargazing is everywhere. It can make people fall in love – as in the Danny Boyle film *The Beach*, as well as in countless other cinematic references. It can cement friendships, as in *The Fisher King*, and it can help turn little girls into good citizens, according to the Brownies, who now have a stargazing badge. It as also often considered to indicate a calm and reflective character.

The Brownies'
stargazing badge

The stargazing scene in *The Beach* has always bothered me. Not because I don't think stargazing has enormous romantic potential but because of the practicalities of the scene. Virginie Ledoyen's character, Françoise, sets up a very expensive camera on the beach so as to be able to take long-exposure photographs of the stars. She points it at the pole so that she gets circular star trails (of the circumpolar stars), then counts while Richard, played by Leonardo DiCaprio, comments in voice-over about this hobby being the sort of thing that could cause you both to fall in love with someone and also later to break up. First of all, would she really have been carrying this telescopic camera and tripod (not an especially light combination) with her as she backpacked across Asia, especially to a beach where the sand and sea air would have probably caused it all kinds of damage? Second, even if she had travelled with this expensive 'kit', what would be the point of taking the pictures the night before they swam across to an island when they had already decided not to take their backpacks with them? Where was she planning to develop the picture? I like the film – but really!

The champagne astronomer

If we don't dwell on the flaws, the fact remains that films draw attention to stargazing's huge romantic potential, and August is a great time to exploit that potential. Though the nights are still not particularly dark in much of the northern hemisphere, they are beginning to get longer. Astronomical twilight starts late, which provides an excuse to stay out. What's more, August has one of the most impressive meteor showers of the year to look forward to, the Perseids. While I'll say much more about meteor showers in Chapter 8 – to coincide with the other big annual meteor shower, the Leonids – the Perseids are worth a digression here.

At its peak, a meteor shower appears as just that, a shower of meteors or shooting stars falling from the sky. The name of each shower comes from the constellation the meteors appear to be shooting from, in this case the constellation Perseus. Because Perseus is a northern hemisphere constellation the Perseids can be seen only in the northern sky. Its peak is around 12 August, though the exact date varies slightly from year to year. While the sight itself is far from guaranteed – predictions of the exact date of the peak can often be out by a day or so, while, as with all astronomy, you're very much at the mercy of the weather – it does offer the perfect excuse for a romantic night-time picnic.

But why stop at meteor showers? Provided you have some champagne, some nice, preferably warming food, and enough blankets and cushions, you can invent any number of excursions to look at various aspects of the night sky.

Obviously you need to get away from the glare of the city lights, but where you go will depend on your budget and preference. My boyfriend and I once took a jeep into the Sahara in the early hours of the morning when on holiday in Morocco. We had breakfast under the stars and watched them slowly disappear as the Sun came up.

I understand that taking a trip out away from the bright lights of the coast on a yacht for the night can also make for a very romantic picnic. Alternatively, though this depends very much on personal temperament, there's always camping. One summer in my childhood my family and I went canoeing down the Ardèche River in the south of France. Each night we would drag our canoes on to the bank, get out our sleeping bags and sleep under the stars. The romance of stargazing is all about getting back to nature, about being and feeling a small part of a large universe. Sleeping with nothing between you and a clear, dark sky full of stars can certainly do that for you.

Summer festivals, though often not conducive to genuine romance, can be romantic for the stargazer in the broader sense. At their best, festivals celebrate a return to nature and community. They're about large numbers of people getting along because they have a common interest – be it in music, the environment or litera-ture. They're about bracing the elements together, and at night they're about spending time under the stars together. Over the last couple of years the Institute of Physics has taken telescopes to some of the smaller festivals around the UK, inviting people to take a look. This has been very popular with festival-goers.

The August sky offers more than meteor showers and festivals – for one half of the planet it's too cold for festivals and the meteors are not visible. What they do have in the southern hemi-sphere in August is lots of dark nights just perfect for stargazing. Already we've met various southern constellations, some created by cultures that long pre-date the arrival of Europeans, and others, those astronomers regard as official, that were created by Euro-peans. It's important to remember that the latter were not all created together. There were stages to the process, and at each stage a whole range of constellations were created, some of which were kept, others were kept for a while, and still others were dropped almost as soon as they arrived.

Lacaille's cat

Abbé Nicolas Louis de Lacaille was a French astronomer who, like many men of science of his day, began his studies training for the Church. Instead of taking up a post as a clergyman, on graduation he quickly moved into science, finding work through the Astronomer Royal from France's Royal Observatory in Paris. Lacaille worked as a surveyor, surveying first a part of the French coast, then later the 'French arc of the meridian'.

Surveying was a major project throughout Europe in the eighteenth century. Maps had existed before but were of limited accuracy. Likewise, the art of surveying was not new but it had been used only on a small scale, by landowners surveying their estates rather than countrywide surveys of how different areas linked up. The various European wars of the seventeenth and eighteenth centuries had made it clear to monarchs and governments that more detailed maps were needed, not least to protect vulnerable coastlines and borders. Improvements in surveying techniques and instrumentation made this possible, and this in turn meant that there was plenty of work for astronomers and surveyors in this field. Interestingly, there were many similarities between astronomical and surveying apparatus at this time. Both were about measuring angular heights and distances to establish accurately where things were – whether on Earth or in the night sky. Because of these similarities, often the same people could, and would, engage in both astronomy and surveying. Lacaille was one of those people.

Lacaille's 'arc of the meridian' is the very accurate measure of a very long north–south line or meridian, taking in the curvature of that line and so the curvature of the Earth at that point. Lacaille started by measuring the arc of the meridian between Nantes and Bayonne and followed this up with another in the Cape of Good Hope (the name then given to the whole of the Cape Province in South Africa). His South African 'arc of the meridian' measurements were not very accurate and led him to conclude that the Earth was a sort of pear shape. Luckily, this mistake was corrected by later surveys.

Measuring the arc of the meridian was only a part of Lacaille's mission in South Africa. The main purpose of his trip there was to observe and map the stars of the southern hemisphere. Lacaille chose the Cape of Good Hope for his southern hemisphere research because it is almost directly below France geographically and so would provide a good comparison for the northern hemisphere maps he was familiar with. It was also at the time a Dutch colony, friendly to the French with many French Huguenot (Protestant) refugees. In four years he catalogued nearly 10,000 stars. It is almost as much for the impressive quantity of the work he produced as for its quality that he is remembered today.

In his time at the Cape, Lacaille not only catalogued the stars, but also created new constellations. Where Bayer had gone for the exotic, creating constellations that in some way reflected a European perception of the lands from which they could be seen, Lacaille's were more literal. His new constellations reflected his immediate surroundings as well as celebrating, in a very eighteenth-century way, modern technology.

The illustration overleaf is from Urania's mirror, a set of thirty-two hand-coloured cards, each showing a group of constellations. The stars are marked by holes, with larger holes representing brighter stars, and tissue paper is stuck to the back so that when you hold the card up to the light you see the constellation as it would

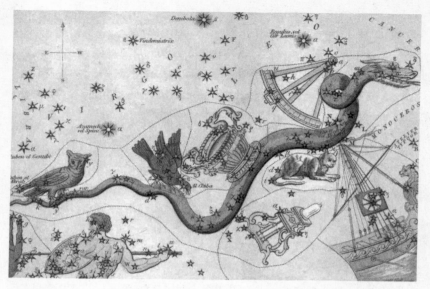

This image of the constellation Hydra surrounded by various smaller constellations, some new, some old and some no longer used, comes from a set of nineteenth-century educational cards called 'Urania's mirror'.

appear in the night sky. A set of these cards formed part of my collection as curator of astronomy at the National Maritime Museum and was always very popular with visitors, especially school groups.

If we look in the middle of the card above we can see Hydra with Corvus the crow and Crater the cup on his back, all constellations we met in Chapter 2. Around them, however, are many new constellations. Just below Hydra's head, on the other side from Sextans, is a cat, Felis, and an iconic piece of eighteenth-century technology, the Air Pump, Antlia Pneumatica.

Lacaille's Felis has not survived the test of time (the same can be said for the owl, Noctua, towards the end of Hydra's tail). In fact, Felis was created by Lacaille as a joke:

> I love cats very much. I will have this picture engraved on the star map. The starry sky has made me tired enough all my life to allow me to have a little fun now.

His other constellations were more serious. Some were quite personal. For example, he created Mensa, originally Mons Mensa or Table Mountain, as a reminder of his view while at the Cape. He also placed his instruments in the sky following the example set by Hevelius and his sextant, Sextans, a constellation found, as the earlier illustration shows, in the same part of the sky as Lacaille's constellations Felis and Noctua. Lacaille's instruments took the form of Circinus, the draftsman's compasses, Norma the set square and Reticulum the recticle (an instrument for measuring the position of stars).

The best way to find these constellations is to begin with the area of the sky we looked at in the last chapter. If you find the Large Magellanic Cloud in Dorado, just above it, towards Chamaeleon, you will find Mensa. It's a very faint constellation but, if you can make it out, you'll see it takes the form of a small curve with one end very close to the Large Magellanic Cloud. Circinus can be found squeezed in between Triangulum Australe and Centaurus on the Milky Way, a sort of thin, triangle-shaped constellation. Norma is further along the Milky Way. If you follow the Milky Way along from Crux to Circinus and Triangulum Australe, Norma is roughly the next constellation you'll come to. But again, Norma is quite faint. Reticulum, another faint one, is the other side of the Large Magellanic Cloud from Volans which, like Dorado, appears to borders the cloud. The star map should, as ever, help you, but they're a faint group so you will need very good conditions to be able to see with any clarity.

Icons of modern technology

Lacaille's constellations also celebrated modern scientific technology. This was the Age of the Enlightenment, when science and its tools were king. In this spirit Lacaille created the constellation Fornax, the chemical furnace, and Telescopium the telescope. Horologium commemorated the sixteenth-century invention of the pendulum clock. Octans, originally Octans Hadleianus or Hadley's Octant, refers to a single groundbreaking piece of navigational apparatus invented in 1730. Finally, Microscopium celebrates the microscope, invented almost at the same time as the telescope in the early 1600s.

The real icon of Enlightenment technology that Lacaille put in the sky was the Air Pump, Antlia Pneumatica, now shortened simply to Antlia. The air pump, seen in the illustration of Joseph Wright's 1768 painting opposite, was a device for creating a vacuum and investigating its properties. The air was pumped out of a glass sealed container and the effects of the vacuum on the object under investigation observed. This experiment is still used today, showing, for example, a candle going out as the oxygen reduces, or an alarm clock ringing silently as it loses the air needed to help the sound travel. In the eighteenth century sometimes a slightly more bloodthirsty approach was taken to heighten the drama. A popular experiment was the 'animal in an air pump experiment'. This was most famously depicted in Joseph Wright's aforementioned painting.

Joseph Wright's 1768 painting, An Experiment of a Bird in an Air Pump

Joseph Wright

Joseph Wright was an artist, but the circles he mixed in included a large number of men of science, including several members of Birmingham's Lunar Society. The Lunar Society is one of the most famous informal social and scientific groups of the late eighteenth century. Its members included Josiah Wedgwood (famous for his pottery business) and Charles Darwin's grandfather, Erasmus. They were industrialists, the new rich. Their special interest in science was partly because science was seen as instrumental in the Industrial Revolution and partly because, unlike Classics, it could be understood and mastered by the non-aristocratic self-made man.

The Lunar Society would meet once a month on the full Moon, so that everyone's journey home would be well lit. They would discuss scientific ideas, eat and drink very well (it is said a semi-circular chunk of dining table had to be cut away to accommodate Erasmus Darwin's growing stomach) and then go home. Joseph Wright was never a member of the Lunar Society but he did mix in their circles. His *An Experiment of a Bird in an Air Pump* was, unusually for the time, painted without commission. The painting was bought by Benjamin Bates, a doctor. It was the last in a series of group scenes by candlelight. Another, appealing to a similar audience, was A Philosopher Lecturing on the Orrery in which a Lamp is put in Place of the Sun, sometimes just called The Orrery. An orrery is a mechanical model of the solar system. Sometimes they're called planetaria. Like the air pump large versions of these instruments made popular demonstration pieces for travelling lecturers, as depicted by Wright in the painting.

The Orrery and the Air Pump were painted in a style normally reserved for religious paintings. Today they are considered masterpieces, but at the time they raised questions as to whether a scientific demonstration should be presented as inducing the same level of awe as a religious miracle. Of course the new industrialists loved it.

Where in the sky

Unfortunately, like Lacaille's other constellations, these are all rather faint. Antlia, as you can see from the illustration on page 108, from Urania's mirror, is just below Hydra, on the other side to Corvus and Crater. None of its stars has an apparent magnitude less than 4. Fornax (the chemical furnace) is little better. It has only two stars visible to the naked eye, the brighter of which has an apparent magnitude of 3.87 and can be found (just about) near Phoenix. Use your star charts to guide you here. Horologium (the clock) is a little bigger, though no brighter, and can be found just curving round Dorado and Reticulum. Telescopium sits just beneath the ecliptic, a little below Sagittarius. Microscopium, meanwhile, is up by Telescopium. You'll find it, with practice and very good conditions, between the line of constellations Indus, Grus and Telescopium on one side and Capricornus and Sagittarius on the ecliptic on the other.

Tools of the artist

Lacaille invented three more constellations that have survived the test of time: Caelum the engraving tool; Pictor the painter, once Equuleus Pictor, the painter's easel; and Sculptor, once Apparatus Sculptoris, the sculptor's apparatus. The importance of engraving in scientific as well as artistic fields helps to explain Lacaille's inclusion of Caelum in his set of new constellations. The painter's easel and the sculptor's apparatus need a different approach. The England of the eighteenth century was less divided by discipline than we might assume. In France, the arts and sciences were often discussed within the same salon, and educated people typically had some knowledge and appreciation of both. Though the majority of Lacaille's constellations are based on the tools of science, in these three there is at least a nod to an equivalent importance of tools in art.

Caelum and Pictor are next to each other in the night sky, near Dorado. Both are very faint: the brightest star is in Pictor and has an apparent magnitude of 3.30. Sculptor, another very faint

AUGUST

constellation, can be found just north of Phoenix. Though it can sometimes seem infuriating that Lacaille should have gone to the trouble of creating constellations from such faint stars, it should be remembered that at the time the telescope was becoming relatively common as an accessory to any fashionable home. Whereas all the ancient constellations were created for the naked eye, Lacaille's were created with the telescope in mind.

Argo Navis

Besides creating completely new constellations out of some rather faint stars, Lacaille also broke up the very ancient but very large constellation Argo Navis, a southern constellation created by the ancient Greeks. Though no longer visible from much of the northern hemisphere it could be seen from ancient Greece. They saw it as the ship created for Jason and the Argonauts for fetching the Golden Fleece. Lacaille broke it up while still keeping the basic idea. Rather than having one large constellation representing the ship, Lacaille created four smaller constellations for different parts of the ship. So now we have Carina (the keel), Puppis (the poop deck), Pyxis (some kind of nautical box) and Vela (the sails).

Whether you look for the ancient constellation Argo Navis or its more modern constituent parts doesn't really matter – the stars are the same. They can be found more or less on the Milky Way, a little north from Crux. The star map can guide you as to the shapes to look for. The good news is that these constellations contain much brighter stars and are much easier to see than any of the others we've met so far in this chapter.

➤✦

Eta Carina

A single star in the constellation Carina, Eta Carina, is particularly worth seeking out. This star has puzzled astronomers for centuries. It's a variable star, so its brightness changes over time. When Edmund Halley catalogued it in 1677 it had an apparent magnitude of 4; when John Herschel saw it in the 1830s it had become one of the brightest stars in the night sky. By the end of the nineteenth century it was invisible to the naked eye with an apparent magnitude of 8, but then it brightened to around 4 again. These variations didn't happen gradually. The sudden brightening of the star in the mid-nineteenth century is called the Great Eruption. In 1998–99 there was another sudden increase as the star suddenly doubled in brightness.

The variation in brightness of this star has been attributed to a build-up in radiation pressure, though this isn't always the case with variable stars. Stars radiate light and heat (as well as other parts of the electromagnetic spectrum), and a measure of the rate at which this happens is called the star's luminosity. The force with which they do this is called radiation pressure. It works against the force of gravity that pulls the star together.

Eta Carina's variability isn't its only interesting feature. It is also very, very, very big. It is the size (or mass) of about 100–150 of our Suns, which is about as big as a star can be, and produces somewhere in the region of four million times as much light. There are not many stars as big or as luminous as this, and none that has been studied as much. Its size means that it is approaching or possibly exceeding 'Eddington's Limit'. This is the limit for how big a star can be before the radiation pressure pushing the star out exceeds the force of gravity holding it all together. If this limit is exceeded,

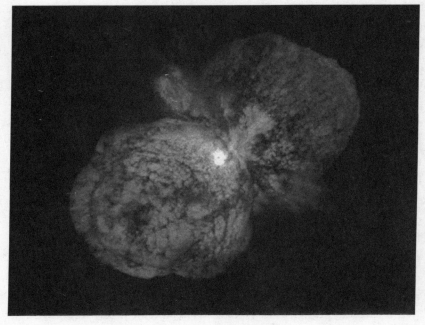

The star Eta Carina close up

AUGUST

the star breaks up. Eddington's Limit gives a theoretical upper limit to the size of stars. Eta Carina is, if not at, then somewhere very near to this limit, and one popular explanation of its odd behaviour over the years is that it has been trying but failing to break up.

Had Eta Carina gone suddenly very bright and then dimmed just once, that would probably have been a sign that it had become a supernova. That it keeps brightening and dimming has made astronomers refer to it as a supernova impostor and its great eruption of the mid-nineteenth century as a supernova impostor event.

A supernova is a star that becomes suddenly very bright and then dims over the course of a few weeks or months. It gets its name from the belief, when they were first observed, that what was being witnessed was the birth of a new, very bright, star rather than the sudden brightening of an existing one. Our Sun will never become a supernova, but many bigger, more massive stars will. When one of these more massive stars stops converting elements into heavier elements in its core it explodes, sending out a shell of material and energy. What's left is a very dense core of neutrons (subatomic particles) called a neutron star. When the original star is very massive, the core left behind is so dense it does not even let light escape; this is called a black hole.

The shell of material and energy that spreads out very suddenly (in a shockwave) in the supernova explosion is made up of all the elements that were made in the core of the star – helium, carbon, neon, oxygen, silicon, iron – as well as what's left of the original hydrogen. This process whereby stars create new elements and then distribute them throughout the universe in supernova explosions is how the elements making up the planets, comets, asteroids and everything else came into existence. It is why, rather romantically, we can all claim to be made from stardust.

As well as producing and distributing the material needed to create life, these supernova explosions are often the catalyst for the formation of new stars. The shockwaves produced by these explosions can, if there is a molecular cloud (the star forming part of a nebula) nearby, trigger the process by which these clouds begin to condense into new stars.

A supernova impostor event, on the other hand, is when a star appears to explode but then survives. It is an event that initially fools astronomers before they establish that the star has not become a supernova but have no satisfactory explanation for what has happened instead. In fact it might not be quite the fraud this term suggests. A popular theory of why Eta Carina's brightness keeps varying is that the Great Eruption was a failed attempt by the star to go supernova. The following variations have just been the star trying to adjust and repair itself.

Rather neatly, Eta Carina is a star that can be described in terms of almost all the astronomical categories we've come across so far. It is a double star, it is a hypergiant (like a red giant only bigger) and it is a star within an open cluster within a nebula. It is also a variable star.

Other variable stars

The amateur astronomer John Goodricke was the first to suggest any kind of explanation for variable stars. Studying the variable star Algol in Perseus in 1782 he discovered the star was not one but two stars orbiting each other, with the periodical eclipse of one by the other (an eclipsing binary), which accounted for the variations in brightness. The Royal Society presented him with their most prestigious award for this work and three years later elected him a member of the Society. Unfortunately, he died, probably of pneumonia, before learning of his election.

Goodricke's explanation for the variable star, Algol, does not account for the variability in all stars, however. Other causes include the star physically swelling and contracting or the star rotating. Some variable stars are very regular, with the rise and fall of their brightness occurring over a number of days and then repeating. These are called Cepheid variables, after the first of its type to be discovered, Delta Cephei, in the constellation Cepheus. Polaris is another example of a Cepheid variable. Others can vary over much longer periods and be much less predictable.

The official southern sky

Before the work of Bayer, Lacaille and all who helped them, the southern sky looked fairly empty to the Europeans. Gradually, over time, as we have seen, these spaces were filled.

In his atlas, Bevis made the distinction between the ancient Greek constellations, which he regarded as well established and unlikely to change, and the less predictable new constellations. As a result he included, as one of his final plates, plate 51, below, showing the southern sky with the Greek constellations only. It is only in the last

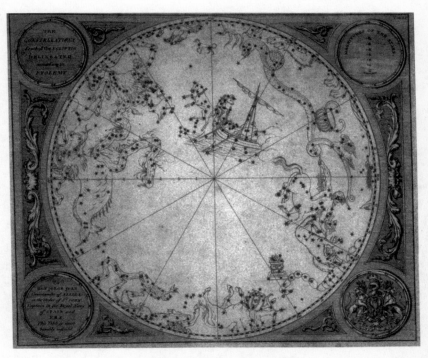

This image from Bevis's atlas shows the southern sky as it was known to the ancient Greeks.

century that we have been able to talk with confidence of southern constellations that have stood the test of time. In 1930 the International Astronomical Union held a meeting in which they set out, once and for all, the official names of all the constellations. They even marked out borders, so that any new stars discovered had to be regarded as one of the official eighty-eight constellations rather than providing an excuse to create a new constellation.

CHAPTER SIX

September and the Milky Way

ACCORDING TO THE ANCIENT GREEKS, the Milky Way is the milk from the goddess Hera's breast. When Hercules was a baby, Zeus thought it would be a good idea to have him suckle at the breast of Hera rather than his mortal mother, so that he would gain godlike strength. Aware that Hera might not be too keen on this plan, he arranged for Hercules to feed only when Hera was asleep. However, one night she woke up and pushed the baby Hercules away from her. The milk that spurted from her breast became Galaxias – from gala, Greek for milk. Our name, Milky Way, is derived from the Latin *Via Lactea*, itself derived from the Greek. As astronomers began to understand what the Milky Way actually was – a very large, gravitationally bound system of stars, dust and gas – they went back to the Greek and called the phenomenon a galaxy.

As we saw in Chapter 4, our galaxy, the Milky Way, is a barred spiral galaxy with our Sun located along one arm.

SEPTEMBER

This is an illustration of what the Milky Way would look like if we were looking upon it from the outside.

Like stars, galaxies are often found in groups, and these groups in clusters. Our local group contains those galaxies we've already met, the Large and Small Magellanic Clouds. It also contains the Andromeda galaxy, which, like the other two, is visible to the naked eye (we'll come back to this galaxy in Chapter 9) and a number of other mostly quite small galaxies. Together with a number of other galaxy groups, these form a galaxy cluster, the Virgo supercluster, where these galaxies are all held together by gravity and revolve around a common centre.

Our Sun is an average-size star, and our galaxy, with around 300 billion stars and an approximate diameter of 85,000 light years, is an average-size galaxy. Like most galaxies it is thought to have a supermassive black hole at its centre, holding it all together. Astronomers have called this Sagittarius A after the constellation in which it is situated.

The Mesopotamians (from modern-day Iraq) created a legend for the Milky Way different from that of the Greeks. Reeds and bark were burned on an altar, Ara, as an offering to the gods, and the smoke this produced became the Milky Way. The offering was made as thanks for the survival of Utnapishtim, the wise king and priest of Shurrupak, after the great floods. (The story is thought to be the origin of the Noah's ark story in the Bible.) The altar became the constellation Ara, often taken to be the start of the Milky Way by those who live in the northern hemisphere and are unable to see the rest of the Milky Way below Ara.

The image of Ara from John Bevis's star atlas shows Ara, upside down, with its flames just touching the feathers of Pavo's tail. The shaded strip on which Ara lies is the Milky Way.

As is so often the case with the ancient Greek constellations, there is more than one story about the origin of the Milky Way, and one of these was adapted from the Mesopotamian myth. Here also, Ara is an altar that produces smoke, which becomes the Milky Way. However, in the Greek version, the altar was made to Zeus' order by one of the Cyclopes, blacksmiths to the gods, and on it a sacrifice was burned. Zeus and his sibling allies hid in the smoke as they attacked the older generation, the Titans. Zeus' side won, restoring order and banishing chaos from the world.

Chocolate manufacturers have helped to make the terms Milky Way and galaxy more instantly recognizable. The Milky Way bar

SEPTEMBER

The constellation Ara

was created in 1923 by an American called Franklin Clarence Mars, who later founded the Mars company. It was the first chocolate bar to be filled with something other than chocolate. Years later, his son named a new chocolate bar 'Mars' after himself and his chocolate-making family. And now we also have the Galaxy bar, as well.

Sadly, for all the myths and the chocolate, it is not easy to see the Milky Way from the northern hemisphere. Our bright cities mean that the Milky Way, though it should be visible to the naked eye, is difficult to see for many of us. From the southern hemisphere it's far easier to find the centre of the Milky Way (around the constellation Sagittarius), which gives a good starting point for finding

SEPTEMBER

the rest. Astronomers sometimes called Sagittarius the 'teapot', because this, as you will see from your star chart, is how this constellation looks from the stars alone. The ancient Greeks may have been able to imagine a centaur drawing a bow, but to modern eyes this formation looks far more like a teapot. For viewers in the northern hemisphere the centre of the Milky Way, when it can be seen at all, is found near the horizon, which puts it nearly, if not totally, out of view.

The Mayan World Tree

The Milky Way, especially for observers in dark skies in the southern hemisphere, can dominate the view. It is therefore not surprising that, beyond ancient Greece and Mesopotamia, myths describing its origin often give it a place of great importance.

The Mayans called the Milky Way the World Tree. This tree can be seen carved in a sarcophagus lid on the tomb of Pakal in the ancient ruins at Palenque, in Mexico. Palenque was a thriving Mayan city before the arrival of the Spanish, complete with various large, pyramid-shaped temples now half surrounded by the encroaching forest. The Mayans saw the World Tree as a path between the underworld (via Earth) and the heavens. The underworld was seen as the part of the Milky Way found below the horizon. This idea of a path between these different worlds is also found in other South and Central American cultures. To the Incas, the Milky Way was a river rather than a tree. The upper world, or heavens, was where the dead went, but also where the living went when dreaming.

SEPTEMBER

⇥⇤

Milky Ways and magpies

In Chinese astronomy the Milky Way was a part of the sky holding various constellations, rather as now in the official descriptions of the sky. But it was also the centrepiece of a story found not just in China but all over South-east Asia, a story that has been retold in various poems and operas ever since. The story tells of Zhi Nu (the star Vega in Lyra) and Niu Lang (Altair in Aquila). Zhi Nu was a fairy, the daughter of the Queen of Heaven, whose job it was to weave beautiful clouds in the sky alongside her sisters. Niu Lang, meanwhile, was a penniless and homeless cowherd, whose magical cow was his only friend and companion.

Confiding in his cow, Niu Lang complained of his loneliness, and so the cow came up with a plan to help him find his soulmate. The cow sent Niu Lang to the river where fairies came from their heavenly palace to bathe. The cow told him to take one of the dresses the fairies had left on the bank. Without her dress the youngest fairy could not fly back to heaven and so had to watch her sisters leave without her. The fairy was, of course, Zhi Nu. When the others had gone, Niu Lang came out of hiding with the dress and asked Zhi Nu to stay with him. She did, staying for several years until her mother, the Queen, noticed her absence. Meanwhile, Niu Lang's magical cow had died of old age but had insisted he should keep her hide for emergencies after she was gone.

Now furious, the Queen came down and snatched Zhi Nu and the lovers' two children and took them all back up to heaven. Recognizing an emergency, Niu Lang grabbed the cowhide, which helped him follow Zhi Nu into heaven. He was just about to reach her when the Queen spotted him, pulled out a hairpin and drew a line between the lovers. The line became the Silver River, the Milky Way. Both

Zhi Nu and Niu Lang were miserable. Eventually the Queen took pity on them, and allowed them to meet, just once a year, on the Silver River. At this time, it is said, all the magpies and crows on Earth gather together to make a bridge across the river to bring the lovers together. This event takes place in mid-August, when the two stars (Vega and Altair for us) are high in the sky and easily visible for stargazers across the northern hemisphere, including China. The choice of black and white birds to make the bridge is presumably a way of explaining the bright and dark regions of the Milky Way that fall between the two stars.

Chinese astronomy

The Chinese constellations stand out from the other alternative constellations we've met so far. In some ways they're quite like our own. They had a long history in which they became the official constellations for a large number of people (as opposed to being adopted only by small cultural groups, as is the case with the Australian Aboriginal constellations, for example). Their development has a well-charted history and, significantly, they were voluntarily given up, as a political gesture in 1912, and replaced by the European ones just a few years before these were made the world's official set.

The traditional Chinese constellations were based not only on mythology but also on the people and institutions around them. They were built up gradually over time, as in the West. A full set of the constellations visible from China became more or less fixed around 310 CE by the star map of astronomer Chen Zhou. After that small changes were made, but this gave the broad framework. Chen Zhou differentiated between the different schools to show clearly the origin of each constellation; but gradually this was dropped. One of the earliest extant Chinese star maps is called the Dunhuang manuscript after the library cave in which it was found. It is now housed at the British Museum and is the subject of an international project to preserve and understand it.

The manuscript is thought to date from around 800 CE. It shows one of the few Chinese constellations to be made up of a similar grouping of stars as in the West. You can see the familiar saucepan shape we know as Ursa Major, here known as Beidou, the northern bushel. The constellations around it make up constellations like Boötes, Ursa Minor and Cassiopeia in the Western system. In the Chinese system constellations tend to be much smaller and more numerous. The constellations surrounding the pole star (the circumpolar constellations) were grouped together to form the central, or purple, palace. Within the palace were constellations named after its inhabitants – an emperor, queen, prince and princess as well as various court officials, servants, rooms, pieces of furniture and an armoury. Three stars in our Hercules, for example, were collectively called Nüchang, meaning 'woman's bed'. In astrological terms (since in China, as in the rest of the world, astronomy and astrology were for a long time interlinked) this represented the ladies of the inner palace and any activity here, such as the passing of a comet or planet, was read as an omen referring directly to the potential future of those ladies.

The rest of the sky was divided into four more palaces, one for each season. The boundaries were provided by the position of the Sun at the solstices and the equinoxes. The ecliptic, as we know from previous chapters, is the area of the sky in which the Sun, Moon and planets appear to move. In Western astronomy this area is divided into twelve zodiac constellations based on where we find

SEPTEMBER

the Sun at different points throughout the year; in Chinese astronomy there are twenty-eight lunar lodges or mansions (Hindu and
Islamic astronomy have both). In the zodiac system Aries marks the
position of the Sun at spring equinox, and so the beginning of
spring; Cancer marks the summer; and so on. Similarly, in the

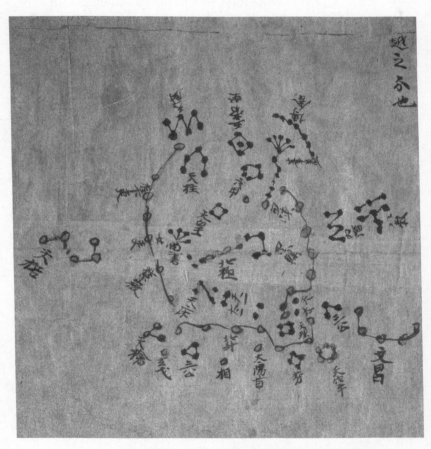

*The Dunhuang manuscript, showing the northernmost constellations
around the pole star*

SEPTEMBER

Chinese lunar lodges, Niao, 'the bird', known to us as Hydra's brightest star, marks the spring equinox; Huo, 'fire', is the brightest star in Scorpius and indicates the coming of summer and the longest day or summer solstice; Xu, 'the void', is one of the brightest stars in Aquarius and marks autumn, while Mao, meaning 'hair', is known to us as the Pleiades in Taurus and marks the shortest day or winter solstice. You may notice that the times of year these stars relate to in the Chinese system are almost the exact opposite to those marked by the Western zodiac. Scorpius here marks the summer solstice, while in the zodiac system it falls nearer to the winter solstice. This is because the Western system is based on the position of the Sun, so what would be visible during the day, while the Chinese system of lunar lodges is based on what is visible at night.

Dunhuang, where the manuscript was found, was once an important stopping point on the Silk Road linking China with the rest of the world. It is curious, given these connections, that China's constellations should have remained so separate. In 1912, with the establishment of the Chinese Republic, the Western system was officially adopted and any separation was abandoned for ever.

Every culture, it seems, has some story to explain the Milky Way. In Scandinavia it has the name 'winter street', as it is used to indicate the coming of winter. To the Khoisan people of Southern Africa the Milky Way is the embers of a fire thrown into the sky by a lonely girl wanting to be allowed to go out and visit people. In various Baltic countries it has traditionally been seen as the pathway of the birds, which could be observed to be using the Milky Way to guide them to warmer, more southerly lands. In Hindu astronomy the Milky Way is known as the Ganges River of the sky . . . and so the list goes on.

Constellations along the Milky Way

For the stargazer, the Milky Way can be looked at simply as a string of constellations. Though really a sort of spiral, with us and our solar system situated on one of its curves, from Earth the Milky Way looks like a circular strip of stars going all the way around us. As such, it crosses the ecliptic (where we find our Sun, planets, Moon and so on) twice.

Starting with Taurus and Gemini we then meet Orion and two constellations we haven't yet met, Auriga and Camelopardalis. Auriga is one of the ancient Greek constellations. A charioteer, he has most popularly been identified with Erichthonius, the mythical inventor of the first four-horse chariot, the quarida. Unlike many ancient Greek constellations, this one does not seem to have originated in Mesopotamian astronomy. It is possible that it was actually invented by the Greeks in recognition of the high value, as a society, they placed on the chariot.

As in the image of Auriga from Bevis's atlas, often Auriga is depicted holding the reins of his chariot (though, as here, the chariot itself is often missing) and carrying goats. 'Goats?' you may ask. The most popular explanation is that the goat or goats were once a separate constellation. Capella, the brightest star in Auriga, is Latin for 'she-goat', while the three less bright stars, ε Aurigae, ζ Aurigae and η Aurigae that mark out Auriga's arm and sit just below Capella, are sometimes known collectively as the Haedi or the Kids. According to myth, the goat Amaltheia acted as wet nurse to the baby Zeus. In recognition of her work she and her two kids were placed in the sky – making their own constellation. Ptolemy brought the constellations together and, despite their rather odd combination, they have remained together ever since.

The constellation Auriga

Farther along is Camelopardalis, which really only just skims the Milky Way. This is the giraffe – a constellation created before both Bayer and Lacaille by Petrus Plancius (see page 91). And then comes Perseus, holding Medusa's head. We will meet this constellation again in Chapter 9, when Perseus comes to rescue the princess Andromeda. Similarly, Cassiopeia and Cepheus, both on the Milky Way, will feature in Chapter 9; we'll save their story until then.

Lacerta and Vulpecula follow. Both constellations were created by the Heveliuses, as was Scutum which we've met already (and

SEPTEMBER

which is also a little farther along on the Milky Way). Lacerta (the lizard) is sometimes described as 'the little Cassiopeia' since they both share the same 'W' shape. Vulpecula (the fox) was originally Vulpecula cum ansere, meaning 'the little Fox with the Goose'. These constellations were later split. Ansere (the goose) was then discarded as a constellation and remains only as a single star – Anser, the brightest star in Vulpecula.

In all, the Heveliuses created seven of the constellations we use today. Their choice of subject appears quite domestic, as though these were simply the objects and animals they saw around them. This certainly holds for Lynx and Lacerta. Though rare now, lynxes, like lizards, were found in Poland where the Heveliuses lived and had their observatory in the seventeenth century.

Cygnus, Sagitta and Aquila, from the Hercules story, are next, and then it's Ophiuchus. This constellation is only just on the Milky Way but is worth mentioning here because of his association with Serpens, the serpent wrapped around his shoulders, whose tail firmly sits on the Milky Way.

Serpens is sometimes described on star charts as two separate constellations, one on either side of Ophiuchus – Serpens Caput (the serpent's head) and Serpens Cauda (the serpent's tail). The brightest star in Serpens is Unukalhai, from the Arabic meaning 'neck of the serpent'. Ophiuchus has various myths associated with him, but a popular one is to identify him as Asclepius, the healer who, by watching serpents, discovered a way of using herbs to keep his patients alive. Fearing he might make the whole human race immortal, Zeus had him killed but placed him in the heavens in recognition of his good work and good intentions.

Next we find ourselves back at the ecliptic and also at the very centre of the Milky Way, with Sagittarius and Scorpius. Norma and

The constellation Ophiuchus and Serpens

Circinus, Lacaille's constellations, come next and then Ara, which we met at the beginning of the chapter, and then Lupus.

Lupus (the wolf), is an ancient Greek constellation, though its name came later, perhaps originating with the Romans, perhaps later still. To the Greeks, Lupus was a more general beast, possibly one killed for the nearby constellation Centaurus, possibly the original Erymanthian boar captured by Hercules in his fourth labour.

Centaurus follows Lupus farther along the Milky Way, then come Crux, Musca and the four constellations Lacaille created from Argo Navis, Carina, Vela, Puppis and Pyxis. Next there is Canis

Major, and then Monoceros. Monoceros was originally Unicornus (and this is how it is depicted in Bevis's star atlas) and is another Plancius constellation like Camelopardalis.

That was quite a list of constellations, but most we've either met already or will look at in more detail in later chapters. To find the others – Auriga, Camelopardalis, Lacerta, Vulpecula, Serpens and with it Ophiuchus, Lupus and Monoceros – the best way to start is to look for the Milky Way and the constellations you already know.

Camelopardalis is a very faint constellation near Polaris, our current northern pole star. Near Camelopardalis is Auriga, which

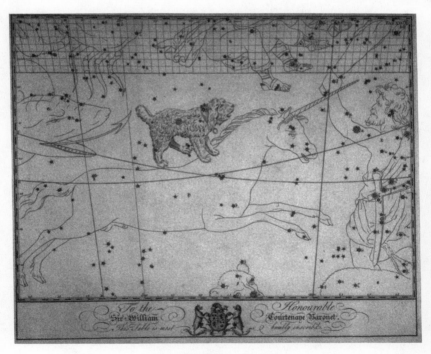

The constellation Unicornis

SEPTEMBER

has the very bright star, Capella, to help you find it. Auriga sits right on the Milky Way and is surrounded by some really bright constellations. In the northern hemisphere in September it can be found between the pole star and the eastern horizon quite late in the evening. Next to it is Perseus and below it Taurus and Orion, all bright constellations and relatively easy to spot. Following the Milky Way west around the pole star you reach Lacerta, another very faint constellation. If you can make it out, it's just to the east of Cygnus. On the other side of Cygnus, sandwiched between Cygnus and Aquila, again very faint, is Vulpecula. Serpens Cauda (the serpent's tail) can be seen towards the western horizon and, with it, Ophiuchus. Both sit just above the ecliptic, while Lupus can be found just below the ecliptic, below Libra. Lupus is definitely easier to see from the southern hemisphere; indeed, for much of the northern hemisphere it never rises above the horizon. Monoceros, meanwhile, can be seen only from the northern hemisphere (at least at a sensible hour) later in the year. This constellation sits near Orion, between his two hunting dogs, Canis Major and Canis Minor, whom we'll meet in the next chapter.

The Milky Way becomes a galaxy

While the various myths surrounding the Milky Way sufficed for a while, there came a time when astronomers began to think about what the Milky Way really was. It had been speculated since antiquity that the glow of the Milky Way consisted of lots of stars, but it took Galileo and his telescope actually to see them, and so show that this was indeed the case. Immanuel Kant, the German Enlightenment philosopher, was, in 1755, the first to suggest the Milky Way was a large spinning disk of stars. He deduced this not

by experiment or observation but through logic and a clear understanding of work that had gone before, including Newton's.

Following Kant, William Herschel is generally credited as the
first person to try to work out the exact shape of the Milky Way
and our position within it. He did this simply by counting stars. He
made a few assumptions, which he stated from the outset. He assumed that all stars produced the same amount of light and so
fainter stars must be further away than bright stars. He also assumed he could see to the very edge of the Milky Way. These
assumptions proved to be false. The shape he came up with was irregular, with a bulge in the centre tapering off to the left and right.
He placed our Sun and solar system towards the middle. We now
know the Sun is about halfway out from the galaxy's core.

Seeing and not seeing the Milky Way

By the nineteenth century astronomers had a clearer idea of what
the Milky Way was, and this knowledge was gradually commu-
nicated to the general population in a deliberate way. Where the
eighteenth century had books and societies aimed at those actively
seeking out astronomical knowledge, by the nineteenth century
many astronomers and well-meaning philanthropists had taken it
upon themselves to extend this market.

In his 1869 book *The Midnight Sky* Edwin Dunkin takes his
reader month by month through the night sky with illustrations
(based on his own calculations) to show what should be visible
from London each month. As you see, the Milky Way was a lot more
visible from London in 1869 than it is today, thanks to modern
street lighting and the all-night lighting of office blocks, shop
windows and billboards.

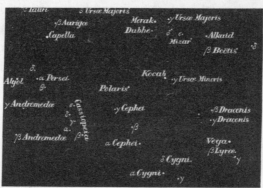

Illustration from Edwin Dunkin's popular science book of 1869, The Midnight Sky, *showing the Milky Way over London*

Dunkin was an employee of the Royal Observatory, Greenwich. His father had been one of the Observatory's poorly paid computers. Dunkin had followed in his footsteps before rising through the ranks to become Chief Assistant, the highest post at the Observatory apart from the Astronomer Royal. His book was published by the Religious Tract Society, a London society founded in 1799 to educate women, children and the working class in a controlled way (they were careful that any scientific conclusions could not be misread as conflicting with, or justifying the rejection of, religious conviction).

Planets in our galaxy

It wasn't until the twentieth century that the Milky Way was discovered not to be the whole universe. Even with the naked eye we can see a few other galaxies outside our own; it was just a question of identifying them scientifically as other galaxies.

While we might now know that the Milky Way is a galaxy and that there are many more galaxies out there, you don't have to go that far to find the possibility of extra-terrestrial life. There are many possible candidates, recently discovered, in our own Milky Way. For extra-terrestrial life, it is generally believed that you need to have another planet not unlike our own. This means you need to have another Sun with a planet orbiting it at roughly the same distance as the Earth orbits our Sun and with certain other shared characteristics. So far nothing conclusive has been found. What we do have are lots of clues and lots of potential clues about what makes our planet so perfect for life and what we would need to look for in other planets to make life there seem likely. These give us some of the main parameters for the search for life on extra-solar planets, or exoplanets.

A star more or less like our Sun is often what astronomers look for. It mustn't be too hot, too big or too variable in the heat and light it gives out. As far as distance goes, from our own solar system we know Venus is too close, Mars is too far away and the Earth is just right. We know this is true because we know that there is life on Earth but not on the other two planets. We also know that we need water and sufficient heat and light from the Sun for there to be life – too close and the oceans evaporate, too far away and the planet becomes too cold. This gives astronomers their 'habitable zone'. There is also the question of planet type. A gaseous giant like Jupiter, Saturn, Uranus or Neptune would be no good. Again we conclude this from the fact that ours is the only planet in our solar system to support life. The search for extra-terrestrial life in this way is just beginning. Before you can start to get particular about planet type and distance, you need to find stars with solar systems.

Before 1781 the universe, as we on Earth knew it, had five planets besides our own – Mercury, Venus, Mars, Jupiter and Saturn. These are all you can see with the naked eye. In 1781 Herschel discovered Uranus; sixty-three years later Neptune was found, followed in 1930 by Pluto. Pluto was regarded as the ninth planet in our solar system until 2006, when it was downgraded to a dwarf planet by the International Astronomical Union. At the same time the two largest asteroids, Ceres and Eris, were upgraded to this new category. It wasn't until the 1990s that any planets outside our solar system were detected.

Even today we don't have telescopes powerful enough to see planets orbiting other stars. Instead, astronomers have a range of techniques for detecting them indirectly, the most common of which is to detect a very subtle 'wobble' of the star being orbited. Gravity is the force exerted by anything that has a mass. The greater the mass, the greater its gravitational pull. The gravitational pull of a sun is what keeps a solar system together and moving round. What is less obvious is that each of the planets also has a (much smaller) gravitational pull, which it exerts on the other planets and on the sun. The larger the planet, the more detectable its gravitational pull on the movement of the star it orbits. It is this movement, caused by the gravitational pull of a very large planet, that astronomers are looking out for. The star should appear to wobble as the planet pulls and orbits at the same time.

A much scaled-down version of this gravitational relationship can be seen in the way our Moon creates tides in our oceans. The Moon orbits the Earth because of the gravitational pull the Earth exerts on it. Because the Moon also has a mass, albeit a much smaller one than the Earth, it too exerts a gravitational pull on the Earth, and this is what makes the oceans move.

Since at least the beginning of the nineteenth century it has been assumed that other stars might plausibly have their own solar systems; but it wasn't until the 1990s that astronomers had the method and technology in place to start making discoveries.

Since 1995, over 250 extra-solar planets have been discovered within our Milky Way – that's 250 planets orbiting a star other than our Sun. There was a brief moment in 1991 when a planet was thought to have been discovered orbiting the Barnard Star in Ophiuchus, but the evidence was shaky. The first real discovery was a Jupiter-sized planet orbiting a star not unlike our Sun in the constellation Pegasus, a discovery credited to a team of astronomers working at the Observatory of Geneva. Whereas historically

SEPTEMBER

we talk about individuals making discoveries helped by a team of uncredited and often anonymous wives, sisters and servants, today everyone gets credit. This often leads to very long lists of names at the beginning of papers, but is a much better reflection of the amount of work involved.

Within three months of this first real discovery two more planets were found orbiting stars in Ursa Major and Virgo, this time by a team working at the University of Berkeley, California. And so it went on through the 1990s and into our current millennium. To begin with, only very big, Jupiter-sized planets were found, since only these had enough gravitational pull to move their star sufficiently for the star's movement to be detected. Technological improvements in detecting this movement have since been made, so that smaller planets can now be found. It is even possible to differentiate the pull of different planets, which means that more than one planet can be detected orbiting the same star and its mass determined. There are other ways of finding exoplanets, though the 'wobble' of the planet's pull on the star is still the most common.

You'll doubtless have noticed that the constellations in which these planets were discovered are not actually found on what we think of as the Milky Way. This is because the Milky Way is more than just this band of stars crossing our sky. It is, instead, most of the stars we see around us. The band simply represents the highest density of stars. Though we talk about the Milky Way as a disk it is not completely flat, not just one single layer of stars, and since our star is within the Milky Way you would expect to see, as we do, stars above and below us also in our galaxy.

We cannot easily see many of these stars we know to have planets with the naked eye, but we can see some. The brightest is probably Pollux or β Geminorium, the second brightest star in Gemini. This star has an apparent magnitude of 1.15, so it's easy to see, though the movement caused by its planet won't be visible. The planet was only discovered in 2006 and, although we obviously can't see it, it adds something to know you're looking not just at a star but at another solar system. These planets do not have interesting names: the one orbiting Pollux, for example, is called Pollux b; the first ever discovered, in 1995, was named 51 Pegasus b. The system is simple, if uninspiring. Planets are given the name of the star they orbit followed by a lower-case letter.

Stars in Cepheus, Draco, Taurus and Eridanus (γ Cephei, ι Draconis, ε Tauri and ε Eridani) are all similarly visible, with apparent magnitudes between 3 and 4.

Incidentally, the numbering and naming systems for stars, though a little mixed up, does follow some logic that is helpful to know when you're trying to find them. Very bright stars tend to have Greek/Roman names, like Polaris and Pollux, or Arabic, like Altair and Deneb. These will also have a symbol indicating their brightness within their particular constellation, a method that extends down to much fainter stars without their own proper names. In this system, introduced by Johann Bayer of the 1603 star atlas Uranographia fame, Greek letters are used to denote brightness – so the brightest is α, followed by the constellation name; the next brightest β, then the constellation name; and so on. Polaris in Ursa Minor is also known as α Ursae Minoris; Pollux is β Geminorum, and so forth. Dimmer stars (which tend not to be visible to the stargazer) are generally referred to by their listing in one well-known catalogue or another.

The Milky Way has fascinated cultures around the world for millennia. Today it is of great interest to astronomers as the home to other solar systems and the possibility of extra-terrestrial life. For the stargazer the Milky Way offers all these different kinds of stories – from the mythical to the historical to the scientific. As we've learnt, all who've seen it have sought an explanation of its purpose, or been inspired by its wonder.

September

※

CHAPTER SEVEN

October and Orion the Hunter

ORION IS PROBABLY THE MOST recognizable of all the constellations in the northern sky after Ursa Major. It is the one we discover as children, as we learn to spot those three stars of his belt and then the stars around it, marking out his shoulders, knees and feet and, farther out, his club and shield. The fact it is a bright constellation helps also: for the northern hemisphere at least, it starts to become visible in the autumn sky as the nights are beginning to draw in. This combination makes for excellent stargazing. The fact that Orion actually gets easier to see the earlier it gets dark – as autumn turns into winter – means it's a great constellation for children as well.

Capturing children's imagination

There are various ways of getting very small children to look up at the night sky. My daughter is just three and we sing 'Twinkle twinkle

OCTOBER

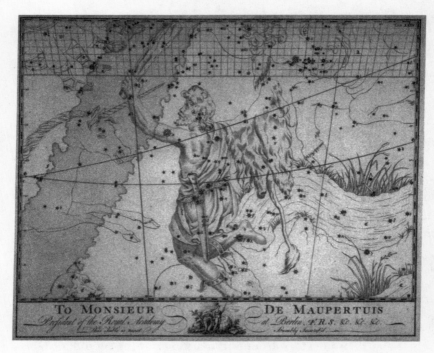

The constellation Orion

little star' to the stars on our way home from nursery. We also look to see if we can spot the Moon. I've been surprised at how much she's taken to both these ruses. We make telescopes (wrapping kitchen roll tubes with tissue paper and decorating them). We even went to a baby planetarium show at the Royal Observatory – with mixed success; she found some of the 3D demonstrations a little too alarming.

There are lots of astronomical books available, and not just seriously educational ones, that are popular with all ages. Three that immediately spring to mind are Oliver Jeffers, *How to Catch a Star*; Simon Bartram, *Man on the Moon*; and Helen Nicolls and Jan

OCTOBER

Pienkowski, *Meg on the Moon*, which are all very much for the pre-school market.

Other junior-stargazing activities include any kind of painting, colouring or cutting and sticking activity – for instance, making your own night sky picture (with lots of glitter) or a star mobile. Older children might find it fun to re-create those constellation cards from Urania's mirror. Make copies of the originals, punch in the holes, have the children colour them in and stick coloured tissue on the back, or have them create the cards from scratch. Another activity might be to create a game of star snap. A couple of popular activities that have already been used at the Royal Observatory include making rockets and re-creating the solar system to scale in Greenwich Park.

'Twinkle twinkle little star'

'Twinkle twinkle little star' has become something of a favourite among stargazing parents. The origin of this nursery rhyme is often misunderstood. Contrary to popular belief, it was not written by Mozart but rather by two sisters at the beginning of the nine-teenth century. The connection to Mozart comes from the fact that he wrote twelve variations on the theme for piano, which pre-date the words. The music was first published, and not by Mozart, as a French rhyme with the title 'Ah! Vous dirai-je Maman' (Ah, shall I tell you, Mother) in 1761. The date should have been evidence enough that Mozart wasn't involved. Instead, it has been incorpo-rated into the myth, demonstrating, so the story goes, that Mozart was clearly a child genius to compose this at the age of five.

The words to 'Twinkle twinkle little star' come from a book of nursery rhymes published by two sisters, Jane and Ann Taylor, in 1806. The words are very much of their time. They demonstrate a new wonder in what stars are as opposed to where they are. This change in focus was crucial to astronomy from the eighteenth to the nineteenth centuries. Jane, the poem's author, wrote extensively throughout her life, though only some of her work was published. She has been compared to her contemporaries Maria Edgeworth and Jane Austen. Her interest in astronomy and the stars came out of a wider interest in nature, as a retreat (so her nephew claimed, at least) from the demands of conversing and acting appropriately within society. Both sisters were taught some astronomy by their father as children, alongside other sciences.

Here is the text of the full poem – though most people only ever sing the first verse:

> Twinkle, twinkle, little star,
> How I wonder what you are!
> Up above the world so high,
> Like a diamond in the sky!
> Twinkle, twinkle, little star,
> How I wonder what you are!
>
> When the blazing sun is gone,
> When he nothing shines upon,
> Then you show your little light,
> Twinkle, twinkle, all the night.
> Twinkle, twinkle, little star,
> How I wonder what you are!
>
> Then the traveller in the dark,
> Thanks you for your tiny spark,
> He could not see which way to go,
> If you did not twinkle so.

OCTOBER

Twinkle, twinkle, little star,
How I wonder what you are!

In the dark blue sky you keep,
And often through my curtains peep,
For you never shut your eye,
Till the sun is in the sky.
Twinkle, twinkle, little star,
How I wonder what you are!

As your bright and tiny spark,
Lights the traveller in the dark,
Though I know not what you are,
Twinkle, twinkle, little star.
Twinkle, twinkle, little star,
How I wonder what you are!

A technique I borrowed from a historical character proved quite successful with school groups visiting the Observatory. Caroline Herschel used it with her nephew, John Herschel, so it has a good pedigree. The idea is simple: show them beautifully illustrated star charts, charts like the Bevis images or those from Urania's Mirror, and ask them to choose their favourite. Caroline found her nephew liked the whale best:

> I had the pleasure of seeing him some part of each day about me, for his Nurse could not please him more than bringing him to aunty's work-room where I used to entertain him with shewing him the constellations in Fl[amsteed's] Atlas of which the Wail was his favourite object, for when the Nurse brought him to me (en cording shrugs) he came with Aunty shew me the Wail!

The constellation Cetus

The Wail, or whale, she refers to is Cetus, which on star charts looks quite fantastic, if not very much like a whale. Its attraction to children is perhaps its very strangeness.

Zodiac constellations

Older children tend to be less interested in the curious and fantastic beasts that might capture the imagination of younger children, and go instead for those with some personal significance. In my experience the ones they really like are the zodiac constellations;

OCTOBER

they like being able to find the constellation that relates to their own star sign. Though astronomers can be a little wary of encouraging this, not wanting to be seen as having anything to do with astrology, there's no denying the genuine appeal these particular constellations have for younger stargazers. Though we'll return to these in the final chapter, it's worth bearing in mind a couple of things about them when it comes to engaging children and young adults. Not all the zodiac constellations are easy to find in the night sky. While constellations like Ursa Major and Orion were created as a means of identifying and talking about a very bright group of stars, the zodiac constellations were created as a means of identifying a section of the sky (one-twelfth of the ecliptic) and marking regular intervals in the year.

The twelve zodiac constellations break up the ecliptic into roughly equal sections. Each section relates to the backdrop of stars we would see behind the Sun (if the Sun weren't so bright) for one-twelfth of the year. They were created in groups of four, or so it's currently thought, in three stages roughly 3,000 years apart. The first four mark the position of the Sun at the two solstices and two equinoxes around 6500 BCE. Precession or the 'wobble of the Earth' means that over thousands of years these astronomical markers shift, so that by 3500 BCE these four constellations no longer marked the solstices and equinoxes and four more constellations were created. Some 3,000 years later a new four were needed.

The story of Orion

Orion was the son of Poseidon, god of the sea and brother of Zeus, and of Euryale, daughter of King Minos of Crete. He was a famous hunter, always out with his dogs Canis Major and Canis Minor, and he was devastatingly good-looking. One day, while visiting the Isle of Chios in the Aegean Sea, just off the coast of Turkey, he fell in love with Merope. Merope's father, King Oenopion, the son of Dionysos, god of wine, was not impressed with Orion as the potential husband of his only daughter. To earn the right to marry her, Orion was to rid the island of all its fierce and dangerous beasts. He succeeded, but instead of being awarded his prize he was just set more tasks. One night, in frustration, he drank heavily of Oenopion's wine. In his drunken state he tried to force himself on Merope; and as punishment Oenopion had him blinded.

Consulting an oracle (possibly the oracle of Delphi), Orion learned that to regain his sight he would need to travel to the most easterly point on the Earth, find the morning Sun and appeal to the Sun god, Helios, to give him back his sight. Helios was, luckily for Orion, sympathetic to his plight. Able to see once more, Orion was set on revenge. On his way to find Oenopion he met his female counterpart, Artemis, the goddess of hunting. Here accounts diverge, though all agree that he upset her or her family in some way. In one account he is too boastful, claiming to be a better hunter than her; in another he tries to rape her; in another he seduces her only to find her brother Apollo disapproves. Whatever the reason for their falling out, the result was that the Earth produced a deadly scorpion under his feet. The scorpion killed him, and the two were transported to the sky, but on opposite sides so that Orion would never again have to face the tiny beast that had finally defeated him.

It's hard to ignore the similarity to the Hercules story from Chapter 2. Our hero is set impossible tasks, he destroys his relationship with the woman he loves while in a state of diminished responsibility and turns to the Oracle for redemption. He even

SPRING STAR CHART: NORTHERN HEMISPHERE

SPRING STAR CHART: SOUTHERN HEMISPHERE

SUMMER STAR CHART: NORTHERN HEMISPHERE

SUMMER STAR CHART: SOUTHERN HEMISPHERE

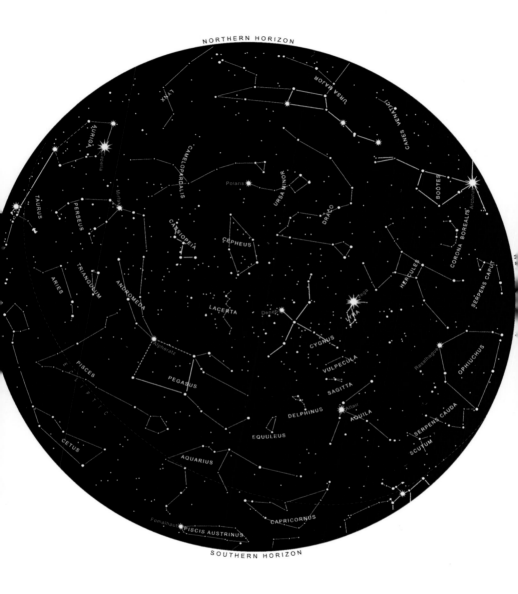

NORTHERN HORIZON

SOUTHERN HORIZON

AUTUMN STAR CHART: NORTHERN HEMISPHERE

SOUTHERN HORIZON

CHAMAELEON
VOLANS
MENSA
CIRCINUS
TRIANGULUM AUSTRALIS
APUS
OCTANS
HYDRUS
RETICULUM
Canopus
ERIDANUS
ARA
PAVO
TUCANA
Achernar
SCORPIUS
NORMA
PHOENIX
TELESCOPIUM
INDUS
GRUS
CORONA AUSTRALIS
Antares
Atnal
SAGGITARIUS
MICROSCOPIUM
SCULPTOR
OPHIUCHUS
SERPENS
PISCIS AUSTRINUS
Fomalhaut
CETUS
CAPRICORNUS
AQUARIUS
ECLIPTIC
AQUILA
PISCES
Altair
EQUULEUS
DELPHINUS
SAGITTA
PEGASUS
VULPECULA
Alpheratz
CYGNUS

WESTERN HORIZON

NORTHERN HORIZON

AUTUMN STAR CHART: SOUTHERN HEMISPHERE

WINTER STAR CHART: NORTHERN HEMISPHERE

WINTER STAR CHART: SOUTHERN HEMISPHERE

wears a lionskin, although this isn't commented on in the Orion story. These similarities have provoked some commentators to suggest that both characters derive from the same source.

The name Orion comes from the Sumerian Uru-anna meaning 'light of heaven'. This is different from the Sumerian name for Hercules, Gilgamesh; so the Sumerians, who pre-date the Greeks and are the source for much of Greek mythology, must have had separate characters too. What is puzzling is that Orion is depicted in the sky wearing a lionskin and fighting Taurus, the bull. No myth appears to exist in which Orion kills a lion or fights a bull; but both the Greek Hercules and the Sumerian Gilgamesh do. The Sumerian Taurus even shares the name of the bull fought by Gilgamesh, Gudanna, or 'bull of heaven'. Given the difficulties of trying to unpick even the most apparently straightforward of myths, this one is likely to remain a puzzle.

Finding Orion and his companions

To find these constellations, the best place to start is with the very familiar – Orion. Find it on your winter star chart and then look up into the sky. In October it rises relatively late in the northern sky, so you'll need to look near the eastern horizon. Exact times will vary according to your location, but, as a rough guide, from London Orion rises at around 10 p.m. (three hours after sunset) towards the end of the month. This gets earlier as you go into November and December, until by about mid-April it is in the sky from twilight until it sets (in the west), around 9 p.m. This gives you October and November to familiarize yourself with the constellation and how to find it before it starts to dominate the early evening sky and becomes the subject of children's stargazing. The constellation

OCTOBER

is most easily spotted by finding the three bright stars, all close together and in a line that makes up Orion's belt. His dogs, Canis Major and Canis Minor, can be found following him, one on either side of the Milky Way, while beneath his feet is Lepus the hare. Taurus can be found just in front of Orion's shield, starting with Taurus's brightest star, Aldebaran.

This set of constellations contains some very bright and well-known stars. In Orion these are Rigel, Betelgeuse and Bellatrix; in Canis Major there is Sirius — and even Canis Minor, which essentially consists of only two stars, has the very bright star Procyon. These are all familiar names, but not perhaps for astronomical reasons. Bellatrix and Sirius are probably better known now as characters in J. K. Rowling's Harry Potter books, while Betelgeuse was the title of a Tim Burton film, albeit with a different spelling.

Bellatrix and Sirius are not the only astronomical names Rowling uses in Harry Potter. Sirius's family is full of astronomical references. Not only are Sirius and his cousin Bellatrix named after stars, but so too are Sirius's brother, Regulus, and his uncle, Alphard. Regulus is in Leo, Alphard in Hydra. Constellations appear as names too. Bellatrix's sister is called Andromeda while her nephew is called Draco (Malfoy). These are all characters from one very dark, though frequently quite glamorous family. What the stargazing fans of Harry Potter are supposed to make of that connection I'm not sure. That two turn out to be good despite their inheritance is surely ground for hope.

More doubles

Sirius and Bellatrix, alongside Betelgeuse, Rigel and Procyon, are interesting to the stargazer for more than just their literary and cinematic connections. For a start, both Sirius and Procyon have names gained from their position within their constellations. Procyon is the brighter of the two main stars that make up Canis Minor. Its name comes from the Greek meaning 'before the dog', in reference to its position following the Dog Star, Sirius, across the night's sky.

With an apparent magnitude of −1.47, Sirius is the brightest star in the whole sky. It has many names, but the best-known are probably the Dog Star, for its place in the constellation Canis Major, and Sirius, from the Greek meaning 'glowing'. To the Egyptians it was known as Sepdet, and, as we have seen, its annual pre-dawn rising (or helical rising, meaning 'with the Sun') was seen as an early warning of the annual flooding of the Nile. To the Egyptians Sepdet was a goddess. She was associated with prosperity and her arrival was celebrated with a festival. Sepdet's father was Osiris, one of the better-known Egyptian gods, and with him she produced a daughter, the planet Venus, often known even now as the morning star.

In Sanskrit Sirius is called Mrgavyadha, meaning 'deer-hunter', and represents Rudra, god of the wind, storms and the hunt. Rudra may be an early form of Shiva, one of the primary gods in the Hindu religion. The association between the characteristics of the god and star may again be connected with the climate brought by that star. In India, the rising of Sirius with the Sun does roughly correspond to the monsoon season, where in some regions there is also a higher risk of fierce storms.

Sirius and Procyon are also both double stars. As we learned in Chapter 1, stars that on closer inspection can be resolved into two stars are called double stars. Some are visual binaries (those that revolve around one another), some are optical binaries (those that just appear to inhabit the same area of space). Both Sirius and Procyon are visual binaries.

In fact, Sirius and Procyon are very similar visual binaries. Both comprise a bright main sequence star, Sirius A and Procyon A respectively, and a dimmer white dwarf star, Sirius B (sometimes nicknamed 'the pup'), and Procyon B revolving around it. Main sequence stars are neither new nor old – they're not just forming but neither have they used up all the fuel at their core – while dwarfs are stars coming towards the end of their lives.

The lifecycle of stars

Both Sirius A and Procyon A are main sequence stars, like our Sun, and of comparable size. The mass of a star is generally described in terms of the mass of our Sun and can be anything from 0.1 for red dwarfs to 150 times the mass of our Sun. Sirius A is only twice as heavy (or to be more technically accurate, massive) or two solar masses; Procyon is about 1.5 solar masses. We talk about masses rather than size because it's unambiguous: it tells you how much material is in the star, while the type of star – whether it's a main sequence star, a giant or a dwarf – should give an indication of its density and volume.

Sirius A and Procyon A are called main sequence because they are at the stage in their lifecycle in which their cores are turning hydrogen into helium by a process called nuclear fusion. This stage corresponds to their position on a graph called the Hertzsprung–

Russell diagram. This details the relationship between a star's temperature, absolute magnitude (or luminosity or real brightness) and stage of evolution.

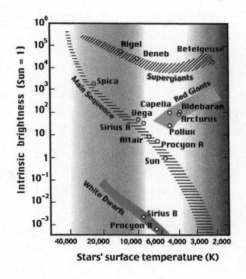

The Hertzsprung–Russell diagram, showing the position of a few well-known stars

The Hertzsprung–Russell diagram tells you where a star is in its lifecycle by measuring its temperature and brightness. It was first drawn up around 1910, before astronomers had a clear idea of how stars evolve. The British astronomer Arthur Eddington (who is probably best known for leading one of the two eclipse expeditions of 1919 that helped confirm Einstein's theory of relativity) used it to better understand the physics behind stellar evolution. Initially, astronomers thought stars might progress steadily through all the different stages of their lifecycle, travelling through the main

OCTOBER

sequence and on to become giants. Eddington showed that, on the contrary, stars tended to stay put on one position within the main sequence for most of their lives: that while they turned hydrogen into helium their temperatures and luminosities stayed more or less constant. Only when they'd finished converting hydrogen to helium at their core did they move off the main sequence to become giants and, finally, in some cases, dwarf stars.

Stars are created when the material in a nebula starts to form distinct groups, to collapse, to be drawn together by gravity. Often this happens when a cloud receives some kind of shock, like the shockwave from a supernova. The collapsed cloud gets hotter and hotter until fusion (the process by which hydrogen is turned into helium) starts to take place in the core. When this happens it is classed as a protostar. The protostar only becomes a real star when its radiation pressure (the force with which it emits heat and light) is as strong as the gravitational force pulling it together. In other words, it has to be balanced. Only when it's balanced can it find a place on the main sequence and so officially become a star. Often a cloud of gas and dust remains around the star in its protostar state. This is called the protostar's protoplanetary disk and can, as the name would suggest, go on to form planets around the star, making a new solar system.

Main sequence stars convert hydrogen to helium. As they start to reach the end of this stage in their lifecycle they start to swell. This is the stage Procyon A is currently at. How fast this happens, and what happens next, depends on how big the star was when it started out. If we go back the Hertzsprung–Russell diagram on page 157, you will see that, above the main sequence line, are giant and supergiant stars. Low mass stars, stars with masses below 0.5 solar masses, never become giants but instead become red dwarfs, one of the smallest type of star. Stars with masses between 0.5 to 6 solar masses go from being main sequence stars to giants, as their outer layers expand. These, as we've seen, go on to form planetary nebulae and finally, as they lose their outer layers completely, dwarf stars. Very high mass stars, meanwhile, become supergiants.

Supergiants are very large stars that, having finished converting hydrogen to helium at their cores, expand to become even larger, hotter, brighter stars. These are the stars most likely to end their lives as supernovas and finally neutron stars or even black holes. Very roughly, the diagram also indicates the relationship between temperature and colour. Very hot stars, stars like Betelgeuse, appear red; relatively cool stars, like Rigel, appear blue.

Below the main sequence stars are dwarf stars. Their position on the diagram shows us that these are relatively dim, cold stars. White dwarfs are thought to be the final stage in the evolution of almost all stars, all except the very massive that instead become black holes.

White dwarfs are usually made up of carbon and oxygen, though they can be made up of heavier elements if the original star was massive enough. The reason white dwarfs are made up of these heavy elements, from carbon upwards, is because of the way stars evolve. In their main sequence stage they fuse hydrogen to helium in their core. As giants they then turn this helium into heavier elements. Once that's done they shed their outer layer and the core, now made up of carbon and oxygen, becomes a white dwarf. Eventually a white dwarf will cool and become a black dwarf, or so it is thought. No black dwarfs are thought to exist just yet since the process takes longer than the current age of the universe. Knowledge of the existence of white dwarfs is a relatively new discovery, only just pre-dating the coining of the term in 1922.

The nature of Orion and the triangles

From the Hertzsprung–Russell diagram you can see that Sirius A and Procyon A are both main sequence stars, as are Vega and Altair from the summer triangle (see page 157). Procyon, Sirius and Betelgeuse form what some astronomers term the winter triangle. Procyon and Sirius also form part of a larger geometric shape, the winter hexagon or winter circle, which takes in Rigel (in Orion), Aldebaran (in Taurus), Capella (in Auriga) and Pollux/Castor (in Gemini) as well.

The winter triangle seems to be a relatively new invention, probably created to help modern amateurs find their way around the night sky. The term was first used at around the same time that the term 'summer triangle' was being popularized by astronomers like Patrick Moore in the mid-twentieth century. The purpose of having these alternative groupings, called asterisms, is to help teach the relationship between constellations in the night sky. While constellations and the stories associated with them help us to remember the patterns of certain groups of stars, asterisms help to show how each group relates to another. In other words, the summer triangle helps us remember where the constellations Cygnus, Aquila and Lyra sit in relation to another. The winter triangle shows us how Orion and his two hunting dogs link up, while the winter hexagon, or circle, links the constellations Orion, Canis Major, Canis Minor, Taurus, Auriga and Gemini for us.

Orion's two brightest stars, Rigel and Betelgeuse, are both supergiants with names deriving from Arabic descriptions of their positions within their constellation, but each is at a slightly different stage in its lifecycle. Betelgeuse is a very large red super-giant, as far as we know one of the largest stars in the night sky. It is a long-period semi-regular variable, meaning its brightness varies unpredictably as it expands and contracts. Ancient Chinese observations of Betelgeuse suggest it might have become a red supergiant between the first century BCE and 150 CE. Chinese observations from the earlier period describe the star as white, while Ptolemy in 150 CE describes it as red. If this is the case and Betelgeuse become a red supergiant only in the first century CE, then it is expected to stay a red giant for some time. Usually stars stay at this stage for tens of thousands of years. At the end of this very long period Betelgeuse is expected eventually to become a supernova, leaving behind a white dwarf. This will not be the standard carbon and oxygen star, however; it will instead be made up of the slightly heavier elements.

Bellatrix, Orion's third brightest star, means 'female warrior'. It is also sometimes called the Amazon star, again referring to its identity with a female warrior. Its Bayer name is γ Orionis, referring to its relative brightness in its constellation. Curiously, Rigel and Betelgeuse's Bayer names do not tally with their actual relative brightness within the constellation. Rigel is the brightest star in Orion, yet Bayer names it β Orionis, while Betelgeuse is the second brightest but is called α Orionis. It's possible their brightness has changed over time, though astronomers dispute this, saying that not enough time has passed for this to have happened.

The Orion nebula

One of the most photographed objects in the sky (excluding objects within our solar system), is the Orion nebula, and, unlike most nebulae we hear about, this one is easily visible with the naked eye.

To the naked eye this nebula is not quite as it appears in this photograph, of course, more as a blurred rather than point-like star. The Orion nebula (M42) is part of a much larger Orion cloud which includes the Horsehead nebula, but it's the easiest to see with the naked eye. It is found on Orion's sword (the three stars below Orion's belt) and is a star-forming nebula or stellar nursery. The Mayans saw it as a smudge rather than a point-like star, but later astronomers catalogued it as just another star until around 1610 when its nebulosity or fuzziness was noted. Stars, and indeed whole

Orion nebula M42

OCTOBER

solar systems, are formed in the Orion nebula. Astronomers have found in this nebula more than just evidence of star formation. They have found protoplanetary disks (from which solar systems are formed) in such large numbers as to suggest that solar systems must be relatively common and those 250 or so discovered to date only a tiny fraction of what's out there.

Once stars form in this, and indeed in any other nebula, they don't just leave. Instead, it is the nebula that evolves. Stars form in little groups within the nebula, these become open clusters, and the remaining nebula is left a little smaller.

Orion the hunter offers a clear way of explaining constellations, the lifecycle of stars and some of the more scientific aspects of the nature of stars. It is the most recognizable constellation, and, for children, obligingly becomes brighter as bed-time draws nearer.

CHAPTER EIGHT

November and Shooting Stars

SHOOTING STARS ARE NOT STARS at all, but meteors; that is, they are little pieces of dust and ice that travel from space into our atmosphere and, on entry, burn up. Every so often, we get what's called a meteor shower, where we see lots of these shooting stars all over the sky (though they appear to originate from the same place) over the course of a few days. Before a scientific explanation of this was worked out, people used to think the stars were literally falling out of the sky. It is not surprising, then, that meteor showers were often seen in astrological terms as omens of doom. Individual shooting stars seem to have fared better, though the idea that it is lucky to wish on a shooting star is probably relatively new.

Meteor showers happen when the Earth, as it travels around the Sun, cuts through the trail left by a comet. Meteorites are very similar, but bigger, bits of material that are too big to burn up completely as they pass through the atmosphere.

In November we have a particularly spectacular meteor shower called the Leonids. The name comes from Leo, the constellation all the shooting stars appear to be coming from. The Leonids appear when the Earth passes through the trail left by the comet Tempel-Tuttle. At their peak, in mid-November, there can be anything up to several thousand meteors an hour apparently radiating from Leo. You need to be somewhere dark, away from street lighting and with a clear, unclouded sky above you to see this impressive sight.

For the northern hemisphere, Leo rises in the early hours of the morning in November, but the meteors themselves should be visible across the sky throughout the night. As a general rule, the

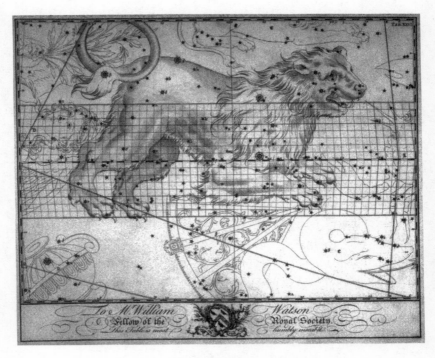

The constellation Leo

NOVEMBER

best view of any meteor shower is when its radiant – the constellation it appears to come from – is high in the sky. In fact, your
chances of seeing thousands of meteors an hour are pretty slim, at
least within the next few years. This is because meteor showers go
in cycles: 1833 saw a meteor storm (where meteor numbers top one
thousand per hour), 1866 saw a similar peak as did 1966, but other
years did not. On a bad year you might see only a dozen or so
meteors per hour – at the shower's peak. Bear this in mind when you
plan your meteor shower watching – the exact peak is very difficult
to predict as is the intensity of the shower at that peak. Having said
that, don't let it put you off. There are so few guarantees in stargazing, not least because it is so dependent on the weather, that you
always need some kind of back-up plan – which is where all those
ideas of romantic, starlit picnics and camping come in. Good food,
maybe something warming like mulled wine or hot chocolate with
brandy in it, tends to help too.

The Great Leonid meteor storm of 1833 was a dramatic sight
to all those caught up in the middle of it. The storm, which reached
its peak in the early hours of the morning of 13 November, was best
seen from America (since it was already light in Europe and anywhere further east). Newspapers reported huge alarm. It appeared
that all the stars were falling out of the sky, and this had to be a message of displeasure from God. The storm contained meteors of all
brightnesses, including very bright ones sometimes termed 'fire-
balls'. Some of these were very bright indeed, lighting up rooms in
which people had been asleep (remember this is in a time before gas
or electric lighting and so no one expected the night to be light).
Few people knew what was really going on, though the storm did
inspire astronomers to find out. After this meteor shower, meteor
science began as a whole new subject.

This depiction of the 1833 meteor storm comes from Adventist Joseph Harvey Waggoner's Bible Readings for the Home Circle, *1888.*

In 1866 and 1867 the Leonids reached another peak. This time, the storm was viewed by a rather better-informed public. Around the same time the Tempel-Tuttle comet was discovered.

Once a comet has been discovered, the calculations begin. This is rarely undertaken by the discoverers but is instead passed over to more mathematically minded astronomers. The Tempel-Tuttle comet was found to have an orbit of thirty-three years. Looking back on past observations, it was clear this was the same comet that had been seen in both 1366 (by Chinese astronomers) and 1699. It was also thought to account for the great meteor storms of 1833 and 1866, and so confident predictions were made for 1899. Unfortunately there was no great storm in 1899. Astronomers made a big thing of the storm's return and hundreds, if not thousands, of people turned out to see it. When the shower turned out to be disappointingly meagre, with only around twenty meteors per hour compared to an expected 100,000 meteors or so the public lost faith in astronomers' predictions. When in 1966 the Leonids did return to their earlier peak, very few people came out to see it.

Speculation about the 1999 storm was encouraged by NASA, which still does not schedule flights during meteor storms, just to be on the safe side. In the event the peak was moderately high – it reached roughly 2,000 meteors per hour (about thirty to forty per minute). The prospects for 2033 will, I'm sure, be the subject of much new speculation, perhaps based on ever-improving models.

Comet Tempel-Tuttle is not the only comet to produce a regular and often spectacular meteor shower here on Earth. The Perseids, which we learned about in Chapter 5, are caused by the passing of the comet Swift-Tuttle. Unlike the Leonids, which can be seen across the sky, the Perseids are mainly visible only from the northern hemisphere. They are also slightly less impressive than the Leonids, tending to reach peaks of around sixty meteors per hour or one a minute. This is enough to make an outing worthwhile. And the time of year helps, since you might feel more inclined to stay out looking for meteors in August than in November.

Two other meteor storms are worth mentioning here, not because they are particularly spectacular in their own right, but because of the significance and fame of their parent comet. The Eta Aquarids and Orionids both result from the passing of Halley's Comet. The Eta Aquarids appear to radiate from the constellation Aquarius and, even more specifically, from a single star – Eta – in this constellation. At its peak, this shower, visible each year from late April to early May, consists of only about a dozen meteors per hour. For the best per-hour rate of meteors, it's best viewed from the southern hemisphere. In the north, Aquarius rises in the morning towards dawn in late April and early May, which reduces the chances of seeing very much. The meteor shower was not officially discovered (though there were previous recorded sightings) until 1870, when it was found by Lieutenant Colonel G. L. Tupman, who was sailing in the Mediterranean. A few years later the relationship between the shower and Halley's Comet was discovered by William Herschel's grandson, Alexander, who had by this time become something of an expert in this relatively new field of meteor science.

The second shower to result from Halley's Comet is also better viewed from the southern hemisphere. The Orionids can be seen annually in late October, appearing to radiate from just above Orion's right shoulder (assuming he's looking down on us), above the star Betelgeuse. At its peak, the shower consists of around thirty meteors per hour, slightly more in the southern hemisphere, slightly less in the northern. The shower was discovered by an American, E. C. Herrick, but it was Alexander Herschel who worked out its radiant – that the meteors all seemed to originate from Orion.

All these showers were discovered, or at least first commented upon in detail and studied, in the nineteenth century. Among those who took an interest in looking for meteor showers were the poets Blake and Tennyson. Alfred Lord Tennyson, England's Poet Laureate from 1850 until his death in 1892, had an observation platform built in his house on the Isle of Wight specifically to see meteors. Similarly, William Blake, best known for writing the words to Britain's alternative national anthem, 'Jerusalem', was fascinated by meteors and meteor showers. His illustrations for Edward Young's 'Night Thoughts' in 1795 contain numerous comets and meteors, even when not specifically mentioned in the text. The illustration he provided for Thomas Gray's poem, 'The Bard. A Pindaric Ode' includes three meteors to accompany Gray's mention in the line:

> Robed in the sable garb of woe,
> With haggard eyes the Poet stood;
> (Loose his beard and hoary hair
> Stream'd, like meteors, to the troubled air).

Through his book illustration work, Blake came into contact with a wide range of disciplines, including astronomy. He became interested in the subject, especially the historical and mythical interpretations of various celestial phenomena. His inclusion of meteors and meteor showers in his illustrations for Young, particularly for the apocalyptic 'Night Thoughts', draws on a historic connection made by the ancients between meteor showers and death, destruction and apocalypse.

Superstitions about meteor showers and comets were hard to shake off. As late as the sixteenth century, seeing comets as omens was still accepted even by those sceptical of some aspects of astrology. Shakespeare, while critical of personal horoscopes in some of his plays, accepted that his characters should regard comets as signposts – as to the fortunes of kings, at least. In *Henry VI, Part I*, for instance is the line:

Hung be the heavens with black, yield day to night!
Comets, importing change of time and states,
Brandish your crystal tresses in the sky,
And with them scourge the bad, revolting stars
That have consented until Henry's death!

While similarly in *Julius Caesar*:

When beggars die there are no comets seen:
The heavens themselves blaze forth the death of princes.

Going back even further, one of the most famous comets suppos-edly predicting bad fortune for a monarch is undoubtedly Halley's Comet, as illustrated in the Bayeux Tapestry. The comet is shown passing overhead shortly before Harold is shot in the eye with an arrow at the Battle of Hastings. It wasn't known as Halley's Comet then, of course. At the time, it was simply seen as a bad omen.

A comet, any comet, is a ball of rock, dust, ice and frozen gases several kilometres in diameter. It orbits the Sun in an elongated ellipse (with the odd exception) travelling fast and close to the Sun at one end of the ellipse, then right out, past most of the planets in the solar system. As it comes close to the Sun, the heat from the Sun causes the frozen material in the comet to vaporize, forming its own little atmosphere called a coma, which streams out behind it taking some of the dust with it and giving the comet its distinctive tail. Some of that dust gets left behind, marking out the path of the comet. It is when the Earth's orbit crosses this path that this dust meets our atmosphere, to produce a meteor shower. A comet does not of course generate light on its own; like most other bodies in our solar system, it is visible to us because it reflects light from the Sun.

The comet's nucleus, that is the 'dirty snowball' solid part as opposed to its coma and tail, is made up of all kinds of different elements and compounds. Comets are thought to have originated, along with the rest of our solar system, in the protoplanetary disk that once surrounded our Sun. Some 4,600 million years ago our Sun was formed, alongside other stars, out of a molecular cloud. Studies of very old meteorites have found elements present that could only have been formed in very old stars, which suggests that it was probably a nearby supernova explosion (or several) that triggered our star's birth. Supernovas, if you remember from

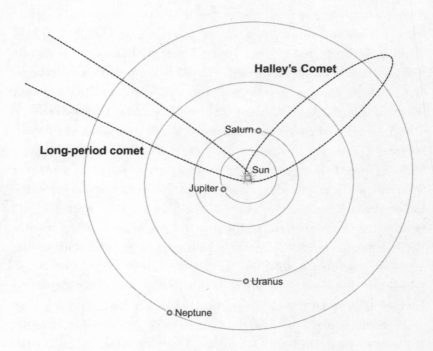

The orbital path of a typical comet

NOVEMBER

Chapter 5, are a stage in the lifecycle of very large stars towards the end of their lives. Having converted hydrogen into progressively heavier and heavier elements in their core by nuclear fusion, they then explode, becoming suddenly very bright, and spreading these elements throughout the surrounding area.

In the very early years of its life, our Sun was surrounded by a disk of dust and gas made up of these various elements distributed by the supernova. Over several million years, this disk gradually formed into the planets, dwarf planets, moons, asteroids and comets we now think of as our solar system. Comets were formed in the outer regions of the solar system where the heat from the Sun was not enough to evaporate away most of the frozen material of which they consist.

Comets exist in one of two places in the outer regions of the solar system: in the Kuiper Belt (or more accurately just outside it in what's called the scattered disk) or the Oort cloud. The Kuiper Belt is the ring of space between the orbits of Neptune and Pluto inhabited by comets and asteroids. Since its demotion to dwarf planet, former planet Pluto is now regarded as being a part of this belt, albeit its largest component. The Oort cloud is even farther out, though still just about regarded as part of our solar system. Comets residing in either one of these regions can be pulled off course by, for example, the gravitational pull of a passing planet inside the solar system or star beyond it. When a comet gets pulled towards the centre of our solar system it goes into its very elongated elliptical orbit, and if the comet is big enough and bright enough, this is when we can see it.

Comets that we all have a good chance of seeing make the news because they're rare; they occur only about once every ten years. Memorable and relatively recent ones include Halley's and

Hale-Bopp's. When and what the next one will be is unclear, not least because new comets are discovered all the time.

Halley's Comet

I remember going out in the street to look for Halley's Comet when I was about eleven, on its much-heralded return in 1986. I also remember the lessons it inspired at school, including one curious one in which the whole class worked – probably for most of the year – on our very own Bayeux Tapestry. No actual tapestry was involved (the school wasn't that confident in our sewing skills) but fabric, wool and sewing were definitely part of the project.

When I first arrived at the Observatory and was led through the collection that I was to look after, a surprisingly large part of it was memorabilia from this 1986 return of the comet. Mostly it consisted of odd things like an alarm clock and some themed bubble gum wrappers – things that nevertheless must have seemed worth collecting on behalf of the nation. This comet's return was an event to be celebrated and commemorated.

One of the main attractions of Halley's Comet is that it can be seen with the naked eye. About once every seventy-five years, its orbit takes it close enough to the Sun and the Earth for us to see it. We see it approach the Sun, it disappears briefly as it goes around the Sun, then we see it again as it leaves the Sun, until eventually it is too far away. Then we need to wait another seventy-five years before we see it again. The American writer Mark Twain was famously born the year Halley's Comet was visible in 1835, and died, as he predicted, on its return in 1910. Like other comets, Halley's is made up of dust and ice and has an orbit that takes it around the Sun, out past Jupiter and back again. It is known as a short-period comet since it returns more frequently than once every two hundred years. When we talk of a comet's return, we refer to its return from the outer reaches of the solar system, not its reappearance from behind the Sun.

Generally speaking, short-period comets come from a different part of the solar system from long-period comets. Short-period comets come from the Kuiper Belt, sometimes called Edgeworth-Kuiper Belt. This is a region of space just outside the orbit of Neptune (since Pluto is now regarded as part of the belt) filled mostly with 'ices', frozen objects made up of substances we tend to know as gases. This belt was first suggested in the 1940s by astronomers Kenneth Essex Edgeworth and Gerard Kuiper, but not actually found until 1992. Just to complicate matters, Halley's Comet doesn't come from there. Rather than destroy the theory, however, scientists have simply concluded that Halley was once a long-period comet which has since speeded up due to the gravitational pull of some of our solar system's larger planets.

But Halley's Comet is unusual for other reasons, too. It is the only comet that can be seen with the naked eye twice in one lifetime. It is also probably the most well-documented comet – certainly there are recorded sightings going back throughout history. Chinese astronomers are thought to have seen it first, recording a sighting in 240 BCE. A few returns later, and sightings were recorded by astronomers throughout the Middle East, in Mesopotamia and Persia. It was seen again in 1066 of course, and when it appeared in 1682 Edmund Halley was inspired to investigate.

Before Edmund Halley each sighting of a comet was thought to be unique. But Halley put together the several sightings at regular intervals preceding the 1682 comet and theorized they might all relate to the same comet. Halley was a friend of Isaac Newton and had been instrumental in getting him to publish his famous book *Principia*, in which, among other things, he sets out his laws of gravitation. By using these laws, Halley worked out the shape and speed of the comet's orbit. Comets travel in an ellipse, or elongated circle. They go very fast around the Sun and then, more slowly, out past the planets and back again. By calculating its orbit, Halley was able to predict the comet's return in 1757–58; sadly he died in 1742 and so didn't see it. The comet did come back when he said, though, and was named after him posthumously, in recognition of his work.

Hale-Bopp

While most short-period comets come from the Kuiper Belt a few, like Halley's, come from the same place as long-period comets – the Oort cloud. Hale-Bopp, which was close enough and bright enough to be visible to the naked eye in 1997, is a long-period comet. The Oort cloud, discovered by one of Kuiper's teachers, Jan Oort, is another region of potential comets surrounding our solar system, like the Kuiper Belt but much farther out. But unlike the Kuiper Belt, this region of space is still just a theory, and has yet to be directly observed.

Hale-Bopp was discovered independently by two Americans. Alan Hale was a semi-professional astronomer when he made his discovery in 1995. After a career in the Navy and a period working for the Jet Propulsion Laboratory (JPL) as an engineer, Hale began a PhD in astronomy. His interest was probably inspired by some of the jobs he was sent on with the JPL, including working on the spacecraft Voyager 2. Having completed his PhD he found the job market for astronomers to be very poor and so he set up the

Earthwise Institute, now known as the Southwest Institute for Space Research. Not long afterwards, he spotted the comet.

Thomas Bopp, meanwhile, was a 'real' amateur. His job had nothing whatever to do with astronomy – he worked as a manager at a construction materials factory – but in his spare time he would look at the stars. Bopp didn't even own his own telescope in 1995, but he did have a good knowledge of the night sky. One night, borrowing a friend's telescope, he saw the comet. (In 1995 you needed a telescope to see it; by mid-1996 and certainly by 1997 it was visible with the naked eye.) He took a photograph and, just as quickly as Alan Hale, made his announcement to the Central Bureau of Astronomical Telegrams, the first port of call for all astronomical discoveries made in America. After checking out the information supplied by both Hale and Bopp, the discovery was confirmed and the comet named after them both.

The comet was spectacular and memorable for some good reasons, and some not so good. It was probably the most observed comet in history, with a huge amount of media coverage guiding people and telling them what to look for and where to look. However, it will also be for ever linked with the mass suicide of the followers of the cult Heaven's Gate. They believed a spacecraft was following the comet (due to a misinterpretation or possible doctoring of one of the photographs of the comet) and that their suicide would somehow allow them to join that spacecraft.

Comet hunters

Hale and Bopp came from a long tradition of amateur and semi-amateur comet hunters. As we saw in Chapter 2, comet hunting began as an eighteenth-century craze when more and more of the middle and upper classes began to own their own telescopes. These telescopes were often beautiful brass and glass instruments that would sit decoratively on a table by a window in a fashionable drawing room or study, acting as much as a conversation piece as a scientific instrument. Of course, some knowledge of the sky was needed, so as not to confuse well-known stars with comets. A list of known fuzzy objects was useful, and Charles Messier created his 1774 catalogue for just that reason – to provide comet hunters with a list of the fuzzy objects.

Messier was a French astronomer and comet hunter. His list of nebulae, star clusters and double stars, objects that, like comets, looked essentially like blurred stars through a telescope but were not, helped cut away the chaff in the hunt for comets. Interestingly, a high proportion of women have been successful comet hunters. Caroline Herschel discovered eight comets, a high number for the late eighteenth century, while Carolyn S. Shoemaker holds the record for the most comets discovered by one person: so far she has discovered a total of thirty-two.

The print of 'The Female Philosopher Smelling out the Comet' dated 1790 is almost certainly supposed to be Caroline Herschel. At the time of her discoveries, a female amateur astronomer, at least one taking credit for her own work, was still something of a novelty. Fanny Burney, the famous diarist and novelist, spoke of her agitation in wanting to view 'the first lady's comet'. The French astronomer and mathematician Jérôme Lalande wrote to Caroline telling her of his plans to have his goddaughter named after her (his godson was already named Isaac after a famous philosopher). Though comet hunting had become relatively commonplace, Caroline's first discoveries still managed to capture the popular imagination precisely because they were a lady's comets.

The woman, described in the title of this print as 'The Female Philosopher Smelling out the Comet' is generally thought to be a caricature of Caroline Herschel.

Reactions to Carolyn S. Shoemaker's comets were more professional. Shoemaker became a professional astronomer relatively late in life, after her children had grown up. Her first job was at Caltech (the California Institute of Technology) where she began to look for asteroids and comets. Over the next twenty or so years she was to discover around eight hundred of the former, thirty-two of the latter. Among her more famous comet discoveries is the Shoemaker-Levy 9, the comet that famously collided with Jupiter in 1994.

NOVEMBER

Fearing the worst

The late 1990s saw a spate of disaster films based on the premise that a comet or meteor could hit the Earth at any moment, wiping out all human life. It's possible that this flurry was partly inspired by the Shoemaker-Levy 9 comet which did, after all, show it was possible for comets to hit planets. Also, the theory that a meteor had wiped out the dinosaurs was rapidly gaining currency. Remember *Deep Impact* and *Armageddon*? Even some astronomers got in on the act with novels based loosely on the worse-case scenario they might come across in their work; Bill Napier's *Nemesis* is one such.

The involvement of professional astronomers in this enterprise hints at something beyond pop culture. With the end of the Cold War, funding for astronomy and space exploration declined rapidly, and astronomers were feeling the pinch. By the late 1990s, what they needed was a political reason for more funding. The idea that the world might end if more money didn't go into the early detection of comets and meteors heading our way seemed a good one for capturing the public imagination. Astronomers have always had to be slightly creative about persuading potential patrons to fund them. The Royal Observatory in Greenwich was founded in 1675 thanks to royal patronage, but only once the king had been persuaded that a very accurate star map was needed to stop cargo ships getting lost at sea. Twenty-first-century astronomers simply have to find reasons that resonate with popular, modern-day concerns.

That's not to say that large meteorites and comets cannot crash land on Earth, causing extensive damage. It is now generally accepted that one, or several, very large meteorites hitting the Earth did have a decisive impact on the fate of the dinosaurs. A huge crater beneath the Yucatán Peninsula in Mexico called the Chicxulub crater dates roughly to the disappearance of the dinosaurs, though it has been suggested other factors were also at play. It's not so much being hit by a meteorite that wipes out life (though obviously nothing trapped underneath is going to survive) as what the meteorite throws up into the atmosphere.

A meteorite is an object that falls to Earth from space. These objects are essentially bits of cast-off material from asteroids that have survived their journey through the Earth's atmosphere. Pieces that burn up completely on their journey through the atmosphere are, as we've seen, called meteors. Asteroids are (more or less) large rocks or very small irregularly shaped planets. The term comes from William Herschel. Perhaps to mark out his discovery of the planet Uranus and see off the competition, Herschel suggested the name asteroid over, for example, the more accurate planetoid, giving the impression that these were more like stars than planets.

When a very large meteorite impacts with the Earth, the force of the hit throws up whatever material is beneath it in a big splash. This puts lots of very small particles into the atmosphere and makes the air very difficult to breathe, blocking out much of the Sun's heat and light and causing climate change. While the moment of impact is very quick, the dust it throws up into the atmosphere takes a long time to settle back down. It is this period of unbreathable air and climate change that is thought to have contributed to the demise of the dinosaurs.

Over the course of history various things smaller than the Chicxulub meteorite have hit the Earth from space. Around 160 craters have been identified on Earth as having come from the impact of some kind of extra-terrestrial object. It's even possible our Moon came from a very early impact. According to this theory, something very large hit the incompletely formed Earth and broke off a piece, which then became the Moon.

Our largest and earliest crater – the Vredefort crater in South Africa – has now been designated a UNESCO site for its special interest to geologists. It is 300 kilometres in diameter and thought to be around two billion years old. Meteor Crater in Flagstaff, Arizona, is one of the best preserved and most visited crater sites in the world. It is on a much smaller scale than the South African crater, which means it's actually easier to comprehend as a meteorite crater. Besides its astronomical and geological importance, Meteor Crater has the added attraction for tourists that it was used to train astronauts for life on the Moon in the 1960s and used in the 1984 film *Starman.*

The most recent meteorite drama took place in 1908 in Tunguska, in Siberia. Early in the morning of 30 June there was a huge explosion, big enough to destroy large parts of a city. According to one estimate, had this happened a few hours later the Earth would have turned enough for it to have completely destroyed St Petersburg. For several nights after the event, nights were very light all over Europe. At first it was thought to have been the result of a straightforward meteorite impact, but when Russian scientists investigated the site in 1927 they found no crater. What later expeditions did find was a huge number of meteorites. Current theories suggest that a very large meteor exploded just above the Earth's surface, scattering small pieces of itself over a large area. The bright nights meanwhile were the result of the dust it left in the atmosphere and that was thrown up on impact by the larger fragments of meteorite.

Given the number of hits our Earth has taken over the years it is prudent to give astronomers support for their careful monitoring of potential impact objects. Already there are a number of research teams around the world dedicated to just that. Astronomers might argue that there should be a few more.

Wishing on a falling star

While the thought of something very big from outer space hitting us can be alarming, we more readily accept little things. Smaller meteorites can be the basis of collections or provide the raw materials for making jewellery; there seems to be a growing market for both.

While seeing a meteorite land is unusual, seeing a shooting star is much less so. And, unlike fears about meteor showers, comets and meteorite landings, shooting stars are considered to be good luck. It's unclear where the superstition for wishing on a shooting star comes from, or when it started, but it has been suggested that it might come from the American nineteenth-century nursery rhyme 'Star Light, Star Bright', which is a happy note on which to leave this star-filled chapter.

> Star light, star bright
> The first star I see tonight
> I wish I may, I wish I might,
> Have the wish I wish tonight.

CHAPTER NINE

December and Queen Cassiopeia

F**OR THE NORTHERN HEMISPHERE**, December is a great time for observing, with its long dark nights surrounding the winter solstice. In the south, nights are at their shortest and brightest and so stargazing must, as it was for the north in June, be largely confined to the Sun. This chapter focuses on a group of constellations that sit very close to the northern pole star, constellations that are therefore very prominent in the December sky in the north.

The princess and the dragon

Once upon a time there lived King Cepheus and Queen Cassiopeia of Joppa, a city in Philistia (Phoenicia, later Palestine). Among their twenty-two children were two daughters, Aeropa and Andromeda. Andromeda was considered especially beautiful. One day, Queen Cassiopeia was overheard boasting about her own and Andromeda's beauty, claiming they were even more beautiful than the sea nymphs, the Nereids. Understandably disgruntled about this, the Nereids complained to Poseidon, their protector. Defending their honour, Poseidon sent floods and a giant sea monster to attack Philistia.

Under attack, King Cepheus went to the Oracle, Ammon, for advice. The Oracle told him the only way to save Philistia was to sacrifice Andromeda to the monster. Andromeda was chained to a rock awaiting her fate when Perseus flew by on the back of his winged horse, Pegasus. Like Hercules and Orion, Perseus is a Greek half-mortal hero and, also like Hercules, the illegitimate son of Zeus. He was just returning home to Greece after having slain the Gorgon Medusa as part of an impossible task set by his mother's jealous admirer, Polydectes. Not only had he come away with Medusa's head, capable of turning all who looked at it to stone, he had also come away with Pegasus, created out of the blood spilled from Medusa's cut throat.

As he flew over the coast of Philistia, Medusa's head in hand, Perseus saw Andromeda chained to the rock and instantly fell in love with her. He offered to kill the sea monster Cetus and save Andromeda in exchange for her hand in marriage. Cepheus and Cassiopeia agreed but then, once Andromeda was safe and Cetus turned to stone (thanks to one look into the eyes of Medusa), they changed their minds. They had people try to disrupt the wedding and kill the bridegroom. But Perseus produced the Gorgon's head and they all turned to stone – after which Perseus and Andromeda lived happily ever after.

Cepheus and Cassiopeia were placed in the heavens by Poseidon, alongside Perseus and Andromeda. Clearly still angry about Cassiopeia's earlier boasts he placed her in a chair near the pole so that for half the year (in the winter) Cassiopeia sits in the sky upside down; that is, she is made up of a group of stars close to the pole star (circumpolar stars) and so is always visible in the northern

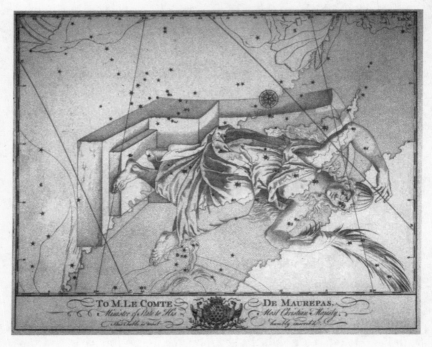

TO M.LE COMTE — Minister of State to His — Most Noble... — DE MAUREPAS — Most Christian Majesty — humbly inscribed

The constellation Cassiopeia

hemisphere but is also always changing orientation. Cepheus meanwhile stands by ineffectually, marked by slightly fainter stars.

Andromeda remains for ever chained to her rock, condemned by her father, with her rescuer, Perseus, complete with Medusa's head and Pegasus, nearby.

Finding the characters

Cassiopeia is a bright constellation, which looks rather like the letter 'W'. It is on the part of the Milky Way nearest to Polaris, the pole star. Next to Cassiopeia, marking out a circle around

Polaris, are Cepheus and then Draco. Draco stretches all the way round to meet Ursa Major on the other side of the pole star. Coming back round to Cassiopeia again, you'll find Andromeda a little way out from the pole star – start to look southward, towards the ecliptic. By her feet is Perseus, who again sits on the Milky Way, just before Auriga and the very bright star Capella. By Andromeda's head is Perseus' winged horse Pegasus. Cetus can be found on the other side of the ecliptic just south of Pisces. It's quite a large constellation and the stars are fairly spread out.

Besides linking up all these clearly visible constellations, this myth also gives us some important clues about the history of Greek constellations. It is generally accepted that much of Greek mythology began with the constellations. But the Greeks didn't make everything up from scratch. Their origins lie, as with so many things, in the Middle East.

Mesopotamia

Though we tend to say the constellations we use today belong to a 'European tradition' of astronomy, or talk of the early constellations as being 'Greek', this is inaccurate. Rather, the development of the constellations and star names we use today has been the product of the work and ideas of astronomers and stargazers passing backwards and forwards across Europe, the Middle East and Asia over centuries, if not millennia.

Until the nineteenth century Europeans believed that the 'Greek' constellations originated entirely with the Greeks. Slowly though, archaeological material from Mesopotamia and in particular the regions of Sumer and Babylon, began to be deciphered. From this it was found that the zodiac constellations were definitely

Mesopotamian in origin, but so too were at least twenty other constellations that have survived the test of time.

The constellations often regarded as ancient Greek were brought together by Claudius Ptolemy, an Egyptian or possibly Greek mathematician and astronomer living in Roman Egypt in 150 CE. In his book Syntaxis (better known by the Latinized version of its Arabic name, Almagest, meaning 'The Greatest'), Ptolemy describes a total of fourty-eight constellations including the twelve zodiac signs. The earliest written description of the twelve zodiac signs comes from a Babylonian text dating from around 400 BCE, so, though pictorial descriptions of them pre-date this, it is thought that these twelve constellations originated in Babylon, gradually building up over the course of a millennium as the apparent path of the Sun, Moon and planets began to be studied more and more.

The Greeks simply imported this zodiac ready-formed, then brought them together and devised a mythology that explained their various shapes and relationships to one another. Later Islamic astronomers took these constellations and combined them with Bedouin stories about the sky. From this combined process came most of our star names, as well as descriptions of where the stars actually are.

Astronomy and religion

Islam, probably more than any other major religion, has a special place for astronomy. Not only is astronomy needed to work out prayer times and key Ramadan dates and times, it is also considered to be the duty of every Muslim to try to better understand the universe created by Allah. This may be why astronomy flourished in the Islamic world.

Last December I was asked to do a radio interview about the star of Bethlehem. Every year at about this time there is a certain amount of interest from the media about the star. Curiously, although Jesus exists in other religions, the star of Bethlehem seems to be linked only to Christianity and the nativity story, where its appearance leads the three wise men to the baby Jesus. Judaism has little to say on Jesus at all, let alone his birth or any star connected to it, regarding him as a historical Jew who has been misinterpreted and wrongly regarded as the son of God. In Islam he has a more important place, but the Qur'an and the Haddiths, Islam's primary religious texts, give a different account of his birth from that found in the Bible, and there is no place there for the star of Bethlehem.

Even in the Bible there is only a very brief mention of the star. In the New Testament, the Gospel of Matthew 2:1 says:

> Now when Jesus was born in Bethlehem of Judaea in the days of Herod the king, behold, there came wise men from the east to Jerusalem, saying, Where is he that is born King of the Jews? For we have seen his star rising in the east, and are come to worship him . . . Then Herod . . . inquired of them diligently what time the star appeared. And he sent them to Bethlehem . . . When they had heard the king, they departed; and, lo, the star, which they saw in the east, went before them, till it came and stood over where the young child was. When they saw the star, they rejoiced with exceeding great joy. And when they were come into the house, they saw the young child with Mary his mother . . .

From this we are given to understand that the star of Bethlehem is an important star that rises at a particular time of night. This has puzzled astronomers for centuries. Part of their interest stems from wanting to understand every part of the story, but it's also to do with astronomy's usefulness in dating historical events. Although the Christian calendar assumes that Jesus was born in 1 CE few people today accept this as a plausible date. Instead, there is an attempt to work out when he was really born, based on known events. For example, Herod, the king who feared prophecies about Jesus and had all boys under the age of two killed, is said to have died shortly after a lunar eclipse. Just as lunar eclipses are relatively straightforward to predict, so it is equally possible to calculate when they happened in the past. The favourite candidate for Herod's lunar eclipse is that of 30 March, 4 BCE, suggesting that Jesus must have been born around two years before that. This gives a rough time frame; but if the star of Bethlehem could be identified, the date could be pinpointed even more precisely.

Candidates for the star of Bethlehem include a supernova, recorded by Chinese and Korean astronomers in 5 BCE, which would have appeared as an exceptionally bright star not normally seen, and Halley's Comet, which would have been visible to the naked eye in 12 BCE. However, a strong argument against this solution to the puzzle is the almost universal associations of comets with predictions of bad news – hardly the mood related in Matthew. Other comets were visible in both 4 BCE and 5 BCE but have also been discounted for the same reason. The supernova noted by Chinese and Korean astronomers occurred in the constellation Capricornus and was visible in March and April of 5 BCE. Though this clearly doesn't tally with the idea of Jesus being born in December, it does fit with other features of the nativity story, namely that the shepherds were out in the fields with their sheep when they heard the news (much more likely in spring than in the middle of winter).

Another interpretation is that the star was not so much bright as astrologically significant. The three wise men are often described as magi, a term that links them to magic, sorcery and, at a pinch, astrology. An astrological explanation would also help to explain why these wise men were looking for signs in the sky and apparently knew how to interpret them when they found them. Various conjunctions of planets (where two planets appear very close together in the sky) have been suggested as possible astrological signs, but for me the most convincing is the conjunction of Venus and Jupiter in Leo on 17 June, 2 BCE. Here we have Venus, the morning star, which Jesus is said in the Bible to have identified himself with, rising with the Sun. The brightness of planets varies according to where they are on their orbit and so how far away they are from the Earth and the Sun. At this moment, that is when Venus and Jupiter are in conjunction in Leo, Venus should have appeared very bright in the sky at apparent magnitude −4.3. Jupiter, the king planet (significant in itself) was also bright, at apparent magnitude −1.8. Since planets reflect only light from the Sun rather than producing their own light, their brightness varies as their elliptical orbits take them close to and then away from the Sun. Coming together, Venus and Jupiter would have appeared even brighter. That this conjunction should have occurred in Leo is also significant since this constellation is, like Jupiter, associated with royalty. So it appears all to add up. The only problem is that this date does not tally with Herod and his lunar eclipse.

Winter festivals

Both the main candidates for the star of Bethlehem appear at the wrong time of year, but then 25 December was never really thought to have been the actual birthday of Jesus. Rather, it was an appropriation of a much earlier winter festival. Most cultures have some kind of winter festival of lights, often around the winter solstice which, in the northern hemisphere, falls around 21 December. The Romans celebrated on 25 December itself in honour of the god of light, Mithras – the festival thought to have been appropriated by Christianity. As Christianity spread, Christmas soon began to displace other winter festivals, including the pagan festival, Yule. This is why many of the traditions we have – Christmas trees and Father Christmas, for example – have little or nothing to do with the nativity story.

Christmas is not the only winter festival celebrated in the northern hemisphere. The Jewish Festival of Lights, Hanukkah, lasts several days and starts on a date that moves around, according to the Christian calendar, but is fixed in the Hebrew calendar. This is a lunisolar calendar – that is, it uses the movements of both the Moon and Sun. Roughly speaking, this means that the length of the months are determined by the orbit of the Moon round the Earth or the cycle of the Moon through its different phases, while the length of the year is based on the orbit of the Earth round the Sun. Despite this, Hanukkah always falls around the winter solstice. Diwali, the festival of lights celebrated by Hindus, Sikhs, Jains and some Buddhists, falls every year around October/November. The festival lasts several days and the date is set by the Moon, specifically the date of the new Moon within the month of Kartika in the Hindu and Bengali calendars.

Though there is no equivalent winter festival in Islam, the start of Ramadan and Eid, the festival marking its end, are both determined by astronomy. In both cases the festivals are marked by the first sighting of the new Moon within a certain month (just like Diwali). However, there is a difference of opinion within the Islamic community about the meaning of 'first sighting'. This brings us back to my introduction to the importance of astronomy in Islam.

Ramadan is set not by the exact date of the new Moon, which can be calculated with great precision, but with the first sighting of a new crescent. Every Ramadan at the Royal Observatory, we used to get calls from Muslims across the country trying to predict this first sighting. We would also be asked the exact times of sunrise and sunset, which would determine when to start and end fasting each day. The main point of contention was, and probably still is, should Ramadan depend on the first sighting anywhere in the world or the first sighting in each country; or should each Muslim have to see the new Moon himself? These questions explained why there were so many calls on us and our software. We could calculate for them the exact time the new crescent would be wide enough to be seen with the naked eye. We could also tell them which countries would have night at that time and therefore be able to see the new crescent. If Mecca could see it on a particular night, that might determine the beginning of Ramadan if a single sighting was all that was needed for the whole world. Alternatively, we could calculate for any given location whether or not (given good weather, which we could not predict) the crescent would be visible that night.

Ramadan and Eid sometimes fall in winter, sometimes in summer. This is because the Islamic calendar is a purely lunar calendar, based entirely on the movement of the Moon, with no solar component to keep it in line with the seasons. This makes it primarily a religious calendar, often used alongside a separate calendar for farming and other seasonal activities. But it is still, you will note, a calendar based on astronomy.

Andromeda's galaxy

There are a number of interesting objects buried within our winter constellations. The most famous is probably the Andromeda galaxy. Described by Al-Sufi as a 'little cloud' and listed in Messier's catalogue of fuzzy things that are not comets as fuzzy thing number 31 or M31, the Andromeda galaxy is one of only three galaxies visible

The constellation Andromeda

to the naked eye – four, if you count the Milky Way, which you cannot see in the same way you can the other galaxy because we live within it (the other two are the Large and Small Magellanic Clouds). The Andromeda galaxy has an apparent magnitude of about 4.5 and can be found within the constellation Andromeda, just to the right of her torso. The image of Andromeda from Bevis's catalogue shows the beautiful Andromeda half clothed and chained between two rocks. The Andromeda galaxy wasn't known at the time Bevis produced his atlas, but it's roughly where the shaded circle overlaps the chain to the right side of Andromeda's body.

We met the Andromeda galaxy briefly in Chapter 6, in discussing how it relates to our own Milky Way – forming part of the same local group and super cluster. Now it's time to look at the galaxy in more detail, in its own terms. The Andromeda galaxy is big, by far the largest galaxy in our local group: it's thought to contain around a trillion (1,000,000,000,000) stars. It's a spiral galaxy and is hurtling towards the Milky Way at around 3,000 kilometres per second. The Milky Way, meanwhile, is similarly speeding towards it with a collision expected in around 2.5 billion years. If and when this happens, the two will combine to form a huge elliptical galaxy.

The first photograph of the Andromeda galaxy was taken in 1887 by the amateur astronomer Isaac Roberts in his own private observatory. At that time our galaxy was thought to be the only one – and the whole universe, at that – so Roberts interpreted the object as a spiral-shaped nebula in which solar systems and planets were being made.

In 1923 Edwin Powell Hubble found a Cepheid variable in the nebula M31. This discovery led to another – the fuzzy object known

The Andromeda galaxy M31, photographed by amateur astronomer and photographer Robert Gendler

as M31 was not a nebula but a separate galaxy. Enter the Andromeda galaxy and a drastic review of our concept of the universe. Ours was no longer the only galaxy, the whole universe. We were clearly dealing with something much, much bigger. Hubble's discovery made us see, for the first time, how large our universe really is.

Cepheid variables, or at least their importance in working out distances, were fairly new discoveries in 1923. The first Cepheid variable, Delta Cephei in the constellation Cepheus, was discovered by the astronomer John Goodricke in 1784. What makes Cepheid variables special among other variable stars (stars whose brightness varies) is that their period of variability is related to their luminosity or, roughly speaking, brightness. A Cepheid variable whose brightness appears to rise and fall over a period of three days will (overall) be eight hundred times as bright as our Sun. A Cepheid variable whose brightness appears to rise and fall over a period of thirty days will be ten thousand times as bright as our Sun. If you can work out the period of variability, which you do simply by observing it very closely over a long period of time, you can work out its overall brightness. Once you have that you can work out how far away it is by comparing how bright it appears from Earth with how bright you know it to be. This relationship between period and brightness was worked out and published by the American astron-omer Henrietta Swan Leavitt in 1912. It worked not only for the Cepheid variable but also any stars related to it, such as those in the same cluster or, as Hubble discovered, in the same galaxy.

Henrietta Swan Leavitt was one of 'Pickering's women', or, as some less supportive astronomers put it, 'Pickering's harem', at Harvard College Observatory. Like Annie Maunder (née Russell) at the Royal Observatory in Greenwich, Henrietta Swan Leavitt was part of a late-nineteenth-century experiment in employing women in astronomical observatories, implemented by the observatory's director, Edward Charles Pickering. Her job, for which she received half the salary given to her male equivalents, was to sort through the observatory's photographic collection, measuring and cataloguing the brightness of stars as they appeared on the photographic plates. In doing so she was able to study and indeed discover many new variable stars. In addition to her official duties, she gradually worked out the relationship between period and brightness or luminosity in Cepheid variables that Hubble was to make such dramatic use of.

We already know about John Goodricke, the original discoverer of the first Cepheid variable, Delta Cephei, from Chapter 5. His main area of study was variable stars and, besides Delta Cephei he also made another discovery in another of the constellations central to the Andromeda story – that the star Algol in Perseus was an eclipsing variable. Algol's name comes from the Arabic for 'demon', a reference to its position within the head of Medusa that Perseus is seen holding in his hand.

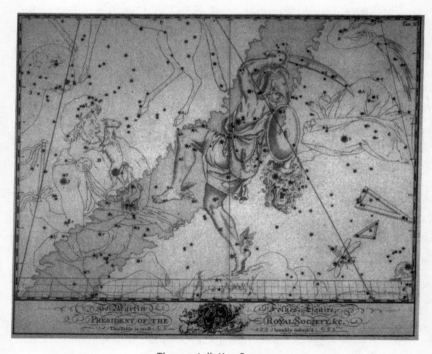

The constellation Perseus

DECEMBER

Algol is the second brightest star in Perseus, with an apparent magnitude of around 2 which suddenly drops to around 3.5 roughly every three days. Though Goodricke was the first to give an explanation for it, this star's variability was known for a long time before that and gained Algol the nickname the 'winking demon'. As you'll remember, Goodricke's explanation involved two stars – an eclipsing binary, where one star eclipses the other's brightness. The dimmer star, β Persei B, revolves around the brighter one, β Persei A. Every three days (actually every two days twenty-one hours), β Persei B eclipses β Persei A, making it seem less bright.

Tycho's supernova

One more dramatic star from this month's group of constellations is worth a mention, though today it's not much to look at. On 11 November 1572, in the constellation Cassiopeia, the brightest of this month's major constellations, there appeared a supernova. Tycho Brahe, the famous Danish astronomer who first saw and recorded it, described the experience in his book *De Nova Stella* (1573) as follows:

> On the 11th day of November in the evening after sunset, I was contemplating the stars in a clear sky. I noticed that a new and unusual star, surpassing the other stars in brilliancy, was shining almost directly above my head; and since I had, from boyhood, known all the stars of the heavens perfectly, it was quite evident to me that there had never been any star in that place of the sky, even the smallest, to say nothing of a star so conspicuous and bright as this. I was so astonished of this sight that I was not ashamed to doubt the trustworthiness of my own eyes. But when I observed that others, on

having the place pointed out to them, could see that there was really a star there, I had no further doubts. A miracle indeed, one that has never been previously seen before our time, in any age since the beginning of the world.

As we know, the Chinese had been recording this type of event for hundreds of years, but for European astronomers this was something quite new. It was the first supernova they'd ever seen.

Tycho Brahe is something of an eccentric figure in the history of astronomy. He had a metal nose, after losing his own in a duel sparked by a mathematical disagreement. He also had a pet moose who died one day from getting drunk on beer and falling down the stairs (why he was indoors at all is not clear). But mostly Brahe is remembered for his astronomy. Without the aid of a telescope, he was able to produce the most accurate sky map in the world, surpassing even Ulugh Beg's map of the mid-fifteenth century. As the son of a nobleman he had access to the best tutors and had little trouble financing his work; after the discovery of his supernova, he even gained the patronage of King Frederick II of Denmark and Norway. The king provided not only financial backing but an entire island, on which Brahe built two observatories. After the death of Frederick II and the arrival of a less sympathetic monarch, Brahe found a new patron in Rudolf II, the Holy Roman Emperor, for whom he produced astrological charts and predictions.

Tycho Brahe's supernova discovery propelled his career. The reason his observations of this 'new star' seemed so extraordinary was that people at the time had inherited a view of the universe from Aristotle that did not allow for any change in the heavens. According to Aristotle and the Christian Church (which had whole-heartedly adopted Aristotle's model and developed it to incorporate a place for God and Heaven), only things on Earth could change. Beyond the Earth's atmosphere lay the fixed stars, God and Heaven – all of which were perfect and unchanging. As a result, when early European astronomers saw comets and meteors they understood them to be phenomena of the Earth's atmosphere, more like weather than astronomy. Changes in the stars they didn't see or, if they did, they somehow managed to explain away.

When Brahe first saw his supernova (nova meaning new star) he thought he'd imagined it. Then he and other observers thought it must be something within the Earth's atmosphere and not a star at all. But then he made some observations that proved that the star must be a long way away, further away than the Moon and so definitely outside the Earth's atmosphere. He did this using 'parallax' – that is, he looked at the star and its surroundings at two completely different places. A good way of understanding parallax is to try it for yourself on a very small scale. Close one eye and look at something close, then swap eyes. The object looks as if it has shifted. If you look in the same way at something further away, the shift should appear less. If the object is a very long way away you don't see any shift at all. Brahe worked out that his new star had to be outside the Earth's atmosphere when he observed it from two different places and found no shift in the star's position.

Brahe's supernova, now known as SN1572, still fascinates astronomers, though for stargazers it's too faint to be visible to the naked eye, and has been since its brightness dropped in 1574, just two years after its discovery. The remnants of this supernova were discovered in the 1960s by astronomers using a very large telescope at the Mount Palomar Observatory in California. It was later photographed by the internationally owned spacecraft ROSAT. It has also been seen with radio telescopes, a type of telescope invented as a spin-off from the military invention radar used in the Second World War. All these points indicate that it was probably a type 1a supernova; that is, an exploding white dwarf star. White dwarfs, as we know, are what remains of medium-sized stars after their outer layers have dispersed. Occasionally these cores can grow more massive as they take (or accrete) matter from a nearby star. This extra mass can then cause the star to heat up and even explode. If this happens they are called supernovas — which is the case of the star that destroyed our idea that the fixed stars were perfect and unchanging.

DECEMBER

CHAPTER TEN

January, Tea and Stars

SINCE WE ARE NOW REACHING the end of the year, it's time to look at some of the constellations we haven't yet covered, the myths associated with them – curiously, almost all date from the ancients – and their most interesting stars.

Corona Borealis

We start with a very bright, easily identifiable northern hemisphere constellation, Corona Borealis. Corona Borealis, the Northern Crown, is made up of a semicircle of bright stars between Hercules and Boötes. It is quite central in the northern hemisphere around mid-January and consequently relatively easy to find, at least once you've found Boötes.

> Looke how the Crowne, which Ariadne wore
> Vpon her yuory forehead that same day

That Theseus her vnto his bridale bore,
When the bold Centaures made that bloudy fray
With the fierce Lapithes, which did them dismay;
Being now placed in the firmament,
Through the bright heauen doth her beams display,
And is vnto the starres an ornament,
Which round about her moue in order excellent.

This was how Edmund Spenser described it in his epic poem *The Faerie Queene* (1596). Spenser was an Elizabethan poet much admired in his own time but also later by the romantic poets, including Byron and Wordsworth, in the early nineteenth century. His casual reference to the stars in *The Faerie Queene* was not unusual for his time. Shakespeare too, like many of his contemporaries, also made such references, because for them and their audiences in Elizabethan England the stars were more familiar than they are today. The skies were darker and the nights longer since there was little in the way of domestic, let alone street, lighting. There was also a greater acceptance of astrology and so some familiarity with the stars was common.

The story of how Ariadne came by her crown has several versions, but the most commonly used to explain the constellation involves Bacchus, a Minotaur and the hero Theseus. The Minotaur was Ariadne's half-brother, the product of her mother's, Pasiphaë's, seduction by a bull. The Minotaur was a dangerous monster and so the King of Minos (Pasiphaë's husband and Ariadne's father) had a labyrinth built to cage him in. Each year, because of a dispute between the King of Minos and the King of Athens, the monster was fed with seven young men and women from Athens.

One day the hero Theseus, the son of the King of Athens, decided he would try to kill the Minotaur and so volunteered to be one of that year's seven victims. When Theseus met Ariadne, they fell in love and she agreed to help him, providing him with a sword and a spool of thread. Once in the labyrinth he killed the Minotaur with the sword, then used the thread (one end of which he'd attached to the entrance) to find his way out.

Theseus and Ariadne then sailed away together. When Ariadne fell ill they stopped at a nearby island. Ariadne fell asleep and when she woke she found Theseus gone. Distraught, she was comforted by the god Bacchus, who had been visiting the island with friends. They soon fell in love, married and lived happily ever after. But Ariadne was mortal and so, when she died, she left her partner to live out the rest of eternity without her. To keep her memory alive, Bacchus threw her crown of jewels up into the sky, where they became the seven bright stars that make up Corona Borealis.

Corona Australis

The second crown in the night sky is the southern hemisphere's equivalent of Corona Borealis – the constellation Corona Australis, or Southern Crown. Though this is one of the forty-eight constellations listed by Ptolemy, and so considered one of the traditional Greek constellations, it does not seem to have a myth associated with it. Often on star maps, as in the illustration from Bevis's atlas, it is depicted as a wreath under the feet of the centaur Sagittarius.

This group of stars, or bits of it, has been identified and used in the constellations of various other cultures besides the ancient

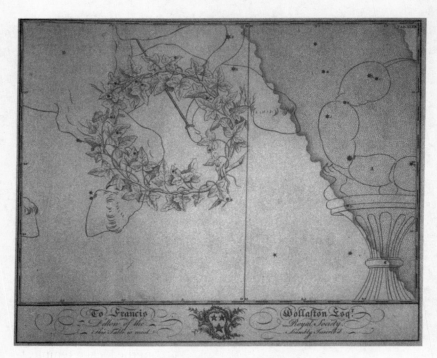

The constellation Corona Australis

Greek. In particular, Al-Sufi mentions several alternative names used by Bedouin astronomers including 'the tortoise', 'the woman's tent' and 'the ostrich's nest'. Sadly, he doesn't seem to have provided the associated mythology, and so why these names were chosen remains a mystery. To Chinese astronomers, the brightest star in this constellation was known as the sixth star of the River Turtle.

Julius Schiller, the German friend of Johann Bayer who produced his own Christianized version of the night sky, also had an alternative name for this group of stars. He called them the Diadem of Solomon, referring to the crown worn by King Solomon, the King of Israel, son of David, and an important figure in the Old Testament.

Coma Berenices

Just on the other side of Boötes from Corona Borealis, near Boötes' brightest star, Arcturus, is the rather faint constellation of Coma Berenices. This constellation comes from Tycho Brahe, who discovered his supernova in 1572. However, unlike other modern creators of constellations, Brahe created a mythology for his, or at least he dipped into existing Greek mythology in his search for a name for his new constellation.

Berenice was the wife of Ptolemy III the King of Egypt in the second century BCE. Following a war against the Assyrians, Berenice shaved off her long golden hair and presented it at the altar of the goddess Aphrodite as thanks for the safe return of her husband. When the hair mysteriously disappeared from the locked temple, the king and queen at first threatened to execute the priests guarding the temple. Luckily for the priests, the astronomer Conon of Samos stepped in, pointing out that her hair was now in the heavens. Aphrodite had been so pleased with the sacrifice, he reasoned, she had immediately transported the queen's golden hair up to the sky.

The characters in this story, with the possible exception of the goddess Aphrodite, were all real historical characters. Ptolemy III really was the King of Egypt in the second century BCE and had a wife called Berenice. He was also known to have been advised by the astronomer Conon of Samos, who is known not only for his place in this story but as a friend of Archimedes.

Delphinus

Because Berenice is known to have been a real, rather than mythical, person, her place in the heavens is unusual. She is not
completely alone, however, thanks to some crafty naming on the
part of certain nineteenth-century astronomers. In the constellation
of Delphinus, the dolphin, are the two stars Rotanev and Sualocin.
Read backwards they spell Nicolaus Venator, the Latin version of
the name Niccolò Cacciatore, an assistant at the Palermo Observatory in Sicily where the first star chart to include names for these
stars was produced. Some say he created the names himself as a
joke, others that his boss, Giuseppe Piazzi, created them in his
honour. Both kept quiet at the time though, since it is not really considered good astronomical etiquette to name stars after yourself.
All those companies claiming you can do this today are selling
something that is not quite what it seems. While they can provide a
certificate announcing the new name for your star, they have no
power to make anyone else use that name. Only the International
Astronomical Union can do that, and they don't approve of such
things. Cacciatore is the only person ever to have managed it.

The constellation Delphinus is a Greek constellation (in that it
was in Ptolemy's list) near the celestial equator; the celestial equator is the ring of space that runs parallel with the Earth's equator
just as the northern pole star, Polaris, is the star directly above the
Earth's North Pole. We observed in Chapter 1 that the Earth is
tilted slightly in relation to the ecliptic. This means that the equator is similarly tilted, and so too is the celestial equator. Although
classified as a northern hemisphere constellation, Delphinus is so
close to the celestial equator that it can be seen from most places
around the world.

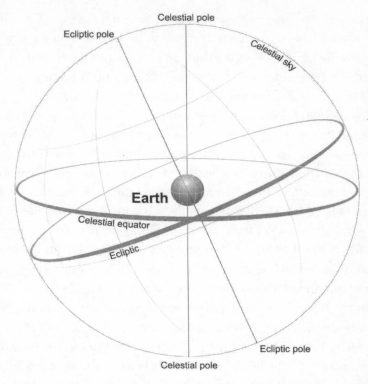

Diagram showing the tilt of Earth and the celestial equator

Delphinus, according to Greek mythology, was the matchmaker who brought together Poseidon, god of the sea, and Amphitrite, one of the sea nymphs or Nereids. Poseidon and the Nereids have of course already featured in some of the constellations we covered in early chapters. It was the Nereids who were so offended by Cassiopeia's claims of superior beauty that Poseidon set the sea monster Cetus on the whole of Philistia (Palestine). Poseidon, so this story goes, had his eye on Amphitrite, but she was not so keen. Fearing his advances she ran away – to the Atlas Mountains in

Morocco, according to some accounts, or simply among her fellow Nereids, in others. Many messengers were sent by Poseidon to try to persuade her to reconsider; but only Delphinus succeeded and so, as thanks, Poseidon found a place for him among the stars.

Though the International Astronomical Union (IAU), and before them generations of globe and mapmakers, have in general favoured Ptolemy's constellation names and later those of European explorers, stargazers have often used pet names for asterisms, or parts of constellations. The brightest stars in Ursa Major are often referred to as the Plough or Saucepan; those in Sagittarius are sometimes called the Teapot while Cassiopeia is often identified more readily as a 'W' than a woman sitting on a chair. Delphinus contains within it a group of stars that mark out a diamond shape and this has led to its rather odd nickname, Job's Coffin. One possible explanation of this is that it relates to a passage in the Book of Job in the Bible: 'Canst thou draw out leviathan with an hook?' and an early depiction of this constellation as a whale or sea monster Leviathan, rather than a dolphin. Despite the biblical associations, however, Job's Coffin is not one of Schiller's constellations. Schiller, you may remember, was the seventeenth-century mapmaker who set out to Christianize the sky by creating alternative names for all the constellations, taken not from the Greeks but from the Bible. For Schiller, Delphinus was 'the water pots of Cana', a name reflecting the pots of water Jesus turns into wine in the New Testament.

Eridanus

The very large southern constellation Eridanus is a river. This mythical river has been identified as many rivers on Earth, from the Nile to the Tigris–Euphrates, from the River Po in Italy to the

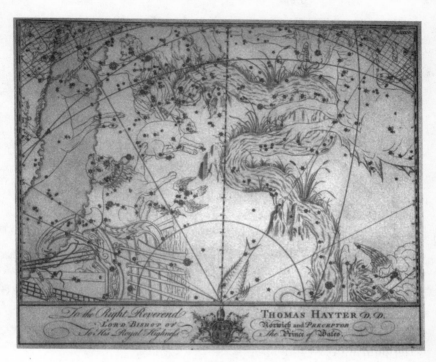

The constellation Eridanus

Rhine and the Rhône. Like the constellation Eridanus depicted in Bevis's atlas, these rivers are all long and winding.

Orion stands at one end of this mythological river, and the river is often mentioned as part of the setting for the Orion story, though it plays no actual part in the Orion myth. Further downstream it meets Cetus and, later still, the Phoenix. The Phoenix would not have been known to the ancient Greeks and so could not have formed any part of the myth. In fact, no other constellations seem to have much to do with the mythology of this constellation. Instead, it stands alone, taking up quite a large part of the southern sky.

JANUARY

Sometimes Eridanus is considered to have formed from the stream of water that flowed from Aquarius the water pourer's pot. This doesn't quite work, though, in terms of how these two constellations are lined up in relation to one another. Preferable is the story of Apollo's son Phaethon, whom we met in Chapter 2. The River Phaethon he crashed into when driving his father's chariot was, in some accounts, Eridanus.

Piscis Austrinus

Like Eridanus and Corona Australis, Piscis Austrinus is another southern hemisphere constellation. The name simply means Southern Fish. This constellation is thought to be the original celestial fish, pre-dating the two fish in the zodiac constellation Pisces, which are sometimes considered to be Piscis Austrinus' children. It contains brighter stars than its better-known offspring and can be found lying on its back with its mouth open under the stream of water pouring out of the pot held by Aquarius.

Though the mythology around this constellation is sketchy at best, there are several accounts that link it to the mermaid goddess of fertility, Derceto, whom the ancient Syrians called Atargatis. Under a spell, Derceto fell in love with a mortal man and bore him a daughter, Semiramis, who went on to become Queen of Babylon. Coming out of her spell Derceto was dismayed at what she'd done. In her panic she killed her husband, abandoned her child and threw herself into the lake at Bambyce (now Manbij) near the River Euphrates. As she entered the water she became a fish and was transported to the heavens as Piscis Austrinis. There are several variations on this story, but all follow more or less this sequence of events. Another possible identification of this constellation in early mythology is with the Assyrian fish god Dagon or the Babylonian fish god Oannes.

Better known than the mythology surrounding this constellation is its single brightest star, Fomalhaut. The name comes from the Arabic, Fum al Hūt, meaning 'the fish's mouth'. Besides its brightness and its unusual name, there is not very much that is remarkable about this star. It is a young main sequence star, about 250 million years old and expected to live to be a billion, so it still has a lot of hydrogen left to convert to helium in its core. It has an apparent magnitude of 1.73, is very bright and, in autumn in the northern hemisphere, it's the brightest star by far, giving it the alternative name, Lonely Star of Autumn. We are, however, now in January, and by January Fomalhaut has become a lot less lonely. In early evening during the early part of January in the northern hemisphere you might just see this constellation, including its bright star Fomalhaut, setting in the west.

Because this star is so bright (the eighteenth brightest in the sky) and, perhaps more importantly, so close – it's only twenty-five light years away – it has been more studied than most. One of the more recent investigations, one carried out using the Hubble Space Telescope, set out to discover whether or not it had its own solar system. What the researchers found were not planets as they expected, but that Fomalhaut had its own Kuiper Belt, or reservoir of comets. Like our Kuiper Belt, which includes comets but also the planet Pluto, this is thought to include not just potential comets but potential planets too. Finding this Kuiper Belt is very unusual, and gives astronomers a great view of what our Kuiper Belt might look like from outside our solar system. The shape of Fomalhaut's suggests there may be planets hidden away between the belt and the star.

Triangulum

Like Corona Australis, this ancient constellation appears to have no place in Greek mythology. It's very small, comprising, as you might expect from the name, just three stars, and is surrounded by the much larger and brighter constellations Perseus, Andromeda and Aries in the northern hemisphere. The only possible myth associated with it is that it is supposed to represent the Island of Sicily, also a kind of triangular shape. It was placed in the heavens by Zeus in response to a request from his sister Demeter, the island's patron, and the goddess of fertility and agriculture.

Demeter and Zeus share another astronomy-related myth – the story of Persephone and the seasons. The seasons, as we know from Chapter 1, are closely linked to astronomy. As a child Persephone lived with her mother, Demeter, away from all the other gods. She grew up to become very beautiful and several of the gods tried to win her; but her mother refused them all and kept her daughter hidden away. One day, Hades, god of the underworld, abducted the young Persephone and took her back to the underworld. Distraught, her mother gave up all her godly duties to search for her. Without her, new seeds on Earth failed to grow and existing plants withered and died. Eventually, Demeter found her daughter and brought her back, but not before Persephone had eaten six pomegranate seeds. The eating of these seeds in the underworld acted as a contract, and Persephone was forever-more obliged to spend six months (one for each seed) in the underworld with Hades. For those six months, as Demeter mourns the loss of her daughter, the Earth is in winter and nothing grows. For the other six months, when mother and daughter are reunited, everything blossoms.

Quasars

In terms of modern, or at least twentieth-century astronomy, the constellation Triangulum has a special place as the home of the first quasar ever discovered. The quasar has been given the name 3C 48 and was discovered in 1960. It is now estimated to be about six billion light years away.

The term quasar is an abbreviation of the full term, quasi-stellar radio source, meaning a radio source that is not exactly a star. As well as light and heat many stars emit radio waves. Light, heat and radio waves all form part of the electromagnetic spectrum. Light, as we know from Newton and his prism experiments, can be broken up into all the different colours of the rainbow. This is just a very small part of the spectrum. Past red at one end there is infrared (related to heat), microwaves and TV and radio waves. On the other side, past violet, are ultraviolet and x-rays. Stars emit electromagnetic radiation as part of the process of transforming atoms at their core. All atoms have energy within them. When one type of atom is converted into another (hydrogen into helium, for example) energy is released and carried away by waves from this electromagnetic spectrum.

Radio astronomy works on similar principles to optical astronomy except that rather than looking at the light waves that travel to us from a distant star, we listen to radio waves. Instead of telescopes with mirrors and lenses, we use dishes (like very large satellite dishes). We can look at and listen to the same things, but we can also listen for things we cannot see. In Chapter 3, we learned about lunar eclipses. During a total lunar eclipse, the Moon looks red because the light from the Sun has to pass first through our atmosphere before it reaches the Moon, and as it does so all the colours except red are scattered away. They do this because they all have shorter wave-lengths than red and this makes them easier to knock off course. This same principle applies to radio and light waves. Radio waves have much longer wavelengths than light and so can travel much far-ther, and past and through more obstacles and so effectively farther than light. This means that by looking, or if you prefer listening, in the radio wave part of the electromagnetic spectrum we see (or hear) things we don't see with optical telescopes.

The first stellar radio source was discovered in the early 1930s quite by accident by an engineer, Karl Guthe Jansky, working for Bell Telephone Laboratories. He had been investigating some inter-ference on certain transatlantic radio phone lines and discovered, through the elimination of all the more likely causes, that the inter-ference was coming from the constellation Sagittarius at the centre of our Milky Way. While Jansky was keen to pursue this, his em-ployers were not, and quickly moved him on to other projects.

Radio astronomy didn't really take off until after the Second World War when the technology, expertise and professional net-works developed during the war's radar programme were put to astronomical use. Bernard Lovell and Charles Husband were the driving force behind the construction of one of the earliest large radio telescopes at Jodrell Bank, just outside Manchester. They built it in 1957 using, in part, material left over from the war. The dish, which looks and is very much like a immensely large satel-lite dish, was first constructed to detect radar echoes of cosmic rays, but these were never found. Instead, the telescope's first dis-covery was radio waves from the Andromeda galaxy. The telescope is now considered one of Britain's best-loved unsung landmarks (at least it won an online competition saying so in 2006).

Since then radio astronomy has become an important part of

regular astronomy. There are now large radio telescopes all over the world, just as there are large optical telescopes. Each has its own advantage. Sometimes stars can appear brighter in radio waves than in light, in which case radio telescopes would be the best way to look at them. Similarly, if a star is behind a cloud of dust we might not be able to see it, but radio waves can pass through dust and so we get another chance.

Lots of celestial bodies emit some radio waves. As Jansky found, our own Milky Way produces some, as does our own Sun: solar flares and sunspots are particularly powerful emitters of radio waves and can sometimes disrupt radio communication on Earth and between the Earth and its satellites and space stations. Radio galaxies are so called because they are particularly powerful emitters of radio waves. Galaxies are thought to have massive black holes at their centre providing a lot of the galaxy's energy and so helping it to emit all these radio waves. Radio galaxies can be detected a very long way away.

It was soon found that there is a limit to how big you can make a single dish and so a single radio telescope. But linking several together creates the same effect as one very large telescope. This is what they've done at the Very Large Array (VLA) group of radio telescopes in Socorro, New Mexico, as well as a number of other radio astronomy observatories. These groups of large dishes make for an impressive sight, and for that reason have been used extensively by film crews. The VLA has been used, for example, in the 1982 film *2010*; the 1985 film *Contact* and the 1996 film *Independence Day*.

Quasar 3C 48 gets its name from its position within the 1959 Cambridge catalogue of radio sources. At that time it was known only to be a radio source. A year later astronomers Allan Sandage and Thomas Matthews found an optical source that matched it, concluding that it was a star, or at least a star-like object that was producing these radio waves. The term quasar was coined a few years later in 1964 by the Chinese-American astrophysicist Hong-Yee Chiu in a press interview. These objects are now thought to be the very massive black holes at the centre of very young (and active) galaxies.

Tea and stars

In the 1830s, the famous astronomer John Herschel (son of William Herschel) and his wife Margaret used to have people round for, as he put it in his diary, 'tea and stars'. Visitors would come to the house, eat and drink tea with the family and then go outside to have a look at what was in the sky that night. It was a very domestic affair. What Herschel's visitors saw in the sky depended not only on what was there but also what he thought they might find interesting. He simply introduced them to constellations that were interesting and visible rather than having some grand astronomical plan to fulfil.

It is difficult today to explain just how popular men and women of science were in the early nineteenth century. Journalists of the day, people like Harriet Martineau, could describe scenes in which 'a group waiting for Sir John Herschel to come out into the street, a rush in the rooms to catch a sight of Faraday . . . ladies sketching Babbage, and Buckland, and Back . . . a train of gazers following at the heels of Whewell or Sedgwick, or any popular artist or author who might be present among the men of science'. Scientists were celebrities, and as such people made a point of trying to get to know them.

When the Herschels moved from England to South Africa for a few years in the 1830s so that John Herschel might catalogue the nebulae, star clusters and double stars of the southern hemisphere, they attracted even more attention. Cashing in on the public's fascination with men of science and their discoveries, a newspaper in America ran a series of hoax articles claiming John Herschel had discovered all kinds of creatures living on the Moon. The articles filled the front page of the *New York Sun* for several days in August 1835, until the story was eventually rumbled. It is now considered one of the most successful and elaborate media hoaxes of all time.

The articles gradually build up the story in instalments. The first says little of the 'discoveries' but makes sure we know how awe-inspiring they are likely to be. 'We are assured', the anonymous author tells us,

> that when the immortal philosopher to whom mankind is indebted for the thrilling wonders now first made known, had at length adjusted his new and stupendous apparatus with the certainty of success, he solemnly paused several hours before he commenced his observations, that he might prepare his own mind for discoveries which he knew would fill the minds of myriads of his fellow-men with astonishment, and secure his name a bright, if not transcendent conjunction with that of his venerable father to all posterity.

The next instalment launched into vivid description of the wonders Herschel discovered and showed his audience the lunar surface:

> The next animal perceived would be classed on earth as a monster. It was of a bluish lead color, about the size of a goat, with a head and beard like him, and a single horn, slightly inclined forward from the perpendicular. The female was destitute of horn and beard, but had

a much longer tail. It was gregarious, and chiefly abounded on the acclivitous glades of the woods. In elegance of symmetry it rivalled the antelope, and like him it seemed an agile sprightly creature, running with great speed, and springing from the green turf with all the unaccountable antics of a young lamb or kitten. This beautiful creature afforded us the most exquisite amusement. The mimicry of its movements upon our white painted canvass was as faithful and luminous as that of animals within a few yards of the camera obscura, when seen pictured upon its tympan. Frequently when attempting to put our fingers upon its beard, it would suddenly bound away into oblivion, as if conscious of our earthly impertinence; but then others would appear, whom we could not prevent from nibbling the herbage, say or do what we would to them.

With his celebrity attracting this kind of weird publicity it is perhaps not surprising that the Herschels preferred a quiet evening with friends and stars to a night out in society. Nor perhaps is it surprising that most visitors to the Cape wanted to be one of those friends, but it was only the chosen few who received an invitation to evenings of 'tea and stars' with Herschel and his family. These visitors were for the most part not astronomers. They were offered a chance to meet the famous astronomer, have a pleasant social evening with him and his family and, almost as an afterthought, see some stars. Today, the amateur astronomy community all over the world runs successful, widely attended and well-organized star parties. It takes a certain amount of serious interest and commitment to stay out all night in a cold field looking at the stars with fellow amateur astronomers.

Wouldn't it be nice to see a revival of tea and stars parties? It seems to me they offer a much gentler alternative, without much need of a schedule. Look up any constellations that might be visible from where you are (and that you know you can spot), find out when they're going to be highest in the sky and then present each of your guests with this information at the start of the evening. You can look these things up relatively easily on the Internet, as well as Moon and planet rising and setting times for your particular location. Alternatively, for the constellations you could use a planisphere, a kind of moveable circular star map for a particular latitude (you can buy them at bookshops). You can then plan dinner around these times, herding everyone out into the garden or street or local common in time to see whatever it is you've chosen to show them.

This could mean some dedicated stargazing or a quick dash outside to see if you can spot the Great Bear, Ursa Major. For the month of January though, perhaps the best choice of stars, at least if you live in the northern hemisphere, might be some of those we've met in this chapter. While Corona Borealis and Corona Australis might not be easy to identify, others are clearer and centrally placed. From London, for example, which is 51.5° North, you might see Delphinus and then Piscis Austrinus setting in the early evening, while Triangulum sits high in the sky. As the night wears on you might see Eridanus near the southern horizon travel slowly from east to west. Then, as midnight approaches, assuming the weather still holds, you should then be able to see Coma Berenices rising in the east. The beauty of doing this in January (in the northern hemisphere at least) is that it gets dark so early you can involve children in the stargazing too and, if you feel so inclined, in the astronomically themed decorating of the house beforehand.

CHAPTER ELEVEN

✴

February, Jason and the Argonauts

As the nights begin to get a little lighter following December's solstice it's time to return once more to the southern sky. Jason and the Argonauts is a well-known story, but from the point of view of the stargazer the main character is not Jason or the Argonauts so much as their ship, Argo Navis, the huge constellation in the southern hemisphere that we looked at in Chapter 5. In Bevis's atlas it is depicted reaching the Clashing Rocks on its way to fetch the Golden Fleece.

Jason and the Argonauts met the Clashing Rocks at the entrance to the Straits of Bosphorus, the boundary between Europe and Asia in modern-day Istanbul. The rocks clashed shut whenever a ship tried to pass through, crushing the ship and all its occupants. However, Jason had been given advice on how to get through by King Phineus, the blind seer, before they set off. Sending a bird in first would cause the rocks to clash shut. As they opened again Argo Navis would have time to row quickly through before the rocks were able to shut again. This is what they did and all, including the bird, survived.

Plate 40 in Bevis's atlas, below, is dedicated to Thomas Whately, a politician and a writer best known for the part he played in the run-up to the American Revolution. As a politician he was involved in the creation of the 1765 Stamp Act, the first attempt by the British Parliament to tax the colonies directly. The tax was an important catalyst for the American Revolution, since Americans felt it had been imposed upon them unfairly. Why Bevis should have selected this very large constellation, prominent at least in the southern sky, to dedicate to Whately is unclear.

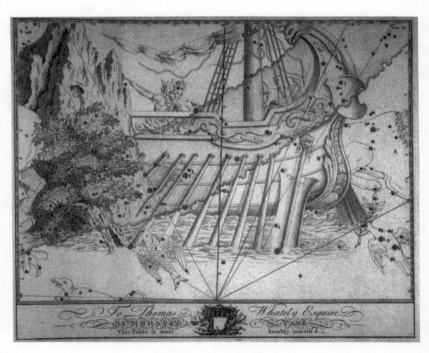

The constellation Argo Navis

Today the constellation Argo Navis, as we have already learned earlier in Chapter 5, has been broken up into four smaller constellations, Carina (the keel), Puppis (the poop or stern), Pyxis (the nautical box sometimes described as the compass) and Vela (the sails). This gives us four of this month's constellations, but the story of Jason and the Argonauts gives us many more, including some we've met already.

Jason and the Argonauts

Jason, our hero, was sent on the seemingly impossible mission to fetch the Golden Fleece by his uncle, the evil Pelias, who had been told by the Oracle that a man fitting Jason's description would be his downfall. If Jason succeeded in his mission, which Pelias was sure he wouldn't, Jason would win back his right to be the King of Iolcus, which Pelias had taken unlawfully from Jason's father. In order to fulfil his seemingly impossible task, Jason brought together a crew for his ship, the *Argo*, made up entirely of heroes. Hercules was one, Perseus another, as were Theseus and Orpheus with his lyre. Polydeuces and Kastor, known to us by their Latin names, Pollux and Castor, the two brightest stars in the constellation Gemini, were also among the party.

The heroes, or Argonauts, set sail, had many adventures along the way, and eventually landed in Colchis on the Black Sea coast. The Golden Fleece belonged to the winged ram Chrysomallos, often identified as Aries, and was now in the possession of King Aeetes of Colchis. Not wanting to give up the Golden Fleece, King Aeetes set Jason three more seemingly impossible tasks. First, he had to plough a field with two fire-breathing oxen. In this he was helped by Aeetes' daughter, Medea, who had been placed under a spell by Aphrodite and fell in love with Jason. She gave Jason a potion to protect him from the fire breathed by the oxen.

Jason's second task was to plant the teeth of a dragon. These grew quickly into soldiers all intent on attacking Jason. Again helped by Medea, Jason followed her advice and threw a stone among them. The soldier Jason hit thought another soldier had thrown the stone, and soon fighting broke out: the soldiers ended up killing one another, leaving Jason alone. Jason's third task was to get past the dragon guarding the fleece. As Medea advised, Orpheus played his lyre and sent the dragon to sleep. Meanwhile, Jason, Orpheus and Medea ran back to the *Argo* with the fleece. In return for her help, Jason promised to marry Medea and love her forever.

Back in Iolcus, after many more adventures on the return journey, Medea used her magic to trick Pelias' daughters into killing him and so freeing the throne for Jason. Disturbed and appalled by her behaviour, the people of Iolcus banished them both and so Jason and Medea travelled to Corinth. There Jason broke his vow to Medea by marrying the princess Glaucis. Medea had her killed and then fled. Jason too was cast out and, after breaking his vow to Medea, fell out of favour with the gods. As he roamed he came across the rotting remains of the *Argo*. Sitting beneath it a part broke away, fell on him and killed him.

With this level of drama, it's hardly surprising the story takes in the four constellations making up Argo Navis, as well as Hercules, Perseus and Lyra (Orpheus' lyre), all of whom we've met already and, on top of that, two zodiac constellations, Aries and Gemini. In addition to all of these is Equuleus, the son or possibly the brother of Pegasus, Perseus' winged horse, which was given to Castor by Hermes (Mercury to the Romans), messenger of the gods and inventor of Orpheus' lyre.

Equuleus, though a small and faint constellation with only a minor part to play in this story, does have one notable claim to fame. It was the location (pinned down sometime after the event) of the famous Great Daylight Meteor Shower of the seventh century. We can still see this meteor shower today. You won't see a shower as such, and certainly not one in daylight, but you might just see the occasional meteor at the shower's peak around 6 February.

Finding those stars

Argo Navis, or Carina, Puppis, Pyxis and Vela, are mainly visible from the southern hemisphere and fill quite a large area of the February sky. They also contain some of the brightest stars in the southern hemisphere, making them relatively easy to spot. If you start with Crux and follow the Milky Way north-eastwards you should come to Vela and Carina and then Puppis and Pyxis as your next bright constellations.

Follow the Milky Way even farther up and eventually, after you've passed Canis Major, Canis Minor and Orion, you reach the ecliptic and, on it, Gemini. Travelling along the ecliptic past Orion again (which is just below it) and Aldebaran, the very bright star in Taurus, you should reach Aries. Of course, to be able to see all this in the sky at the same time requires a certain amount of planning, and will depend on your exact location – you might find that Aries has just set in the west or that Vela has yet to rise in the east.

Neither Hercules, nor Lyra next to him, is likely to be visible in the February sky wherever you are. If you are very far north, you might just see the tips of Hercules' toes peeking over the northern horizon. Equuleus is unlikely to be visible either, at least not at any sensible hour. You might just see it in the late afternoon in early

FEBRUARY

February (it's near Delphinus), but, given how faint it is and how light the sky is likely to be at that time, it seems improbable. Perseus, on the other hand, is still quite central in the northern sky on a February evening. Look out for his bright star Mirfak on the Milky Way, near the even brighter Capella in Auriga.

The stars themselves

Once you've found all these constellations, it's time to look at them a little more closely. In the four constellations that were once Argo Navis there are a couple of interesting stars to look out for. The naming system for the stars in these constellations is slightly odd because of the constellation's history. Whereas most constellations follow the Bayer system, where the brightest stars are named α constellation name, then β constellation name and so on, this is not the case here. Instead, these constellations have retained the old Greek letters relating to their place in their original constellation. Just to confuse matters further, this does not apply to Pyxis. Why Pyxis came to be renumbered when the others were not is something of a mystery.

All the stars in Pyxis are relatively faint; even its brightest, α Pyxidis, has an apparent magnitude of only 3.68. Of the rest, Carina contains the brightest star of them all – Canopus. With its apparent magnitude of –0.7, Canopus is the second brightest star in the night sky after Sirius. If we were closer to it, Canopus would be a very bright star indeed. It is 20,000 times brighter than our Sun but, because it is about 310 light years away, it appears slightly less bright than Sirius. To put this in perspective, Sirius is a mere twenty-two times brighter than our Sun and only manages to outshine Canopus because it is so much nearer.

Canopus is a white supergiant, which is a very massive, hot and luminous star with a relatively short lifespan of only a few million years. We have already met some supergiants – Rigel in Orion is a blue-white supergiant while Betelgeuse in the same constellation is a red supergiant. The colour gives an indication of their temperature and age. Red supergiants are much cooler and older than blue supergiants. Canopus, a white (or yellow-white) supergiant is somewhere between the two.

In Egypt, Canopus was believed to herald the arrival of autumn, as it rose with the Sun at the time of the autumnal equinox. It was also associated with one of the most important Egyptian gods and was known as 'star of Osiris'. Osiris (sometimes identified as the constellation Orion) was the god of life, death and fertility, which perhaps explains the connection with Canopus (autumn brought the harvest).

To Bedouin stargazers, Canopus was Suhail, who unsuccessfully tried to win over the female central one, Al-Jauzah, otherwise known as Orion. His failure to win her heart led to his exile in the far south of the sky. In India, Canopus was known as Agastya, the son of the water goddess Varuna. Agastya was one of the sages or wise men thought by some to have written down the mantras revealed by the Supreme Brahman in the Rig Veda, one of the earliest and most important Hindu texts.

In Carina we also have Eta Carina, or η Carina. This star, as we saw in Chapter 5, is a variable star, probably one that tried and failed to become a supernova in 1843. That year, it outshone all the stars in the sky, including Sirius. Where Sirius would be twenty-two times brighter than our Sun if the two stood side by side and Canopus might be 20,000 times brighter, Eta Carina would outshine them both by being a staggering 4 million times brighter.

Those stars you can see in Puppis and Pyxis with the naked eye are fairly unremarkable, although there is an open cluster, M47, in Puppis that is just about visible. This cluster is actually closer to the constellation Monoceros than Puppis on your map. However, it is classified as part of Puppis because of the boundaries drawn by the IAU in the 1930s to prevent any more constellations being invented to fill in the gaps. As you'll remember, an open cluster is a group of stars that lie close together and were all formed from the same cloud of matter. All stars begin in open clusters before gradually being dispersed. The Pleiades, or seven sisters, in Taurus is the most famous open cluster and we will learn more about this in the next chapter. M47 is much harder to make out and in all but the most perfect of conditions, looks to the naked eye like a single star.

In Vela, the final constellation in the group once making up Argo Navis, the brightest star is γ Velorum. In fact it's not a single star but a multiple star system (like a binary or double star but with more stars). The brightest star within this multiple star system is a Wolf-Rayet star, which is the term for the next evolutionary stage along from a very large supergiant. Charles Wolf and George Rayet were two astronomers working at the Paris Observatory in 1867. Observing the stars in the constellation Cygnus that year they discovered three that displayed notably different spectra from the other stars.

Spectroscopy is a technique developed in the early nineteenth century to find out the composition of a star (or other light source) by looking at the light it emits. As we saw in the last chapter, white light is made up of a spectrum of colours and these form part of the electromagnetic spectrum (which also contains waves that we can't see, such as heat and radio waves). The pioneers of spectroscopy found that when you break up light, from any source, into its spectrum, you tend to find little strips missing. There might be a dark line – a missing bit – in red, another in green, a couple in blue and so on. They tested this on known sources by looking at the spectrum produced from the light burning hydrogen emits, and helium and oxygen and so on. What they found was that each different element produced its own characteristic or signature spectrum. This means that if you look at the spectrum of a star you can work out from what is missing which elements make up that star. This has proved incredibly useful in working out what stars are made of and what is going on inside them at the various different stages in their lifecycle.

In looking at the stars in Cygnus, Wolf and Rayet found a signature they didn't recognize (it turned out to belong to helium, a gas not discovered until a year later). They also found the signatures of carbon, oxygen and nitrogen – none of which is found in most stars. We now know that to have made these, the stars would have had to finish turning hydrogen into helium; and the cores would have needed to be big enough to continue the fusion process so that the helium could then be transformed into these heavier elements. This could only happen in very old, massive stars – in large stars that had already passed the supernova stage. Wolf and Rayet were aware only of the existence of various unusual elements in the stars they were studying in Cygnus. They were not aware of the details of stellar evolution we know about today Despite this, we still use their names to describe these very old and massive stars.

Aries and the equinox

Aries can be seen from around September to March and is best seen somewhere between the two. Like all the zodiac constellations, it disappears from view about the time of year it is thought to have particular astrological significance. This is because, astrologically speaking, the important thing is for the Sun to be in that constellation (this is why they're sometimes called Sun signs), not for us to be able to see it. Aries and all the bright stars within it are associated with the spring equinox and therefore the beginning of the astrological year, though today, due to precession or the wobble of the Earth, the spring equinox actually falls when the Sun is in Pisces.

The stars within Aries have been historically important for their role, at different times, in marking the beginning of spring. Hamal, from the Arabic rās al-ḥamal meaning 'the head of the ram', is the constellation's brightest star and was once the marker for the equinox. Sharatan, the second brightest, and its binary star companion, Mesarthim, both gained their names from their historical place as markers of the spring equinox. Sharatan comes from the Arabic meaning 'the two signs', referring to Sharatan as a binary or double star and to the pair marking the equinox. Mesarthim is thought to have derived either from the Sanskrit name meaning 'the first star of Aries', or from the Arabic meaning 'fat ram' or from the Hebrew meaning 'ministerial servants'. Mesarthim is itself a binary star, making the whole Sharatan–Mesarthim system a multiple star system.

Most of the star names we have come across so far have been Greek or Arabic in origin, or come from some much later European numbering system. This makes a possible Sanskrit or Hebrew origin for Mesarthim unusual. Though alternative names for many stars are sometimes given in Sanskrit or Hebrew, since both languages did at one time have their own systems for naming the stars, very few of these names have found their way into any official listings. For now, the exact origin of Mesarthim remains a mystery.

Castor and Pollux

Castor and Pollux are the bright stars marking out the heads of the two twins in the constellation Gemini. Castor is the brighter of the two, with an apparent magnitude of around 2. Actually Castor is not one star but six – like γ Velorum, it is a multiple star system containing a number of binary pairs. Pollux is slightly less bright.

In mythology, Castor and Pollux are the twins who joined Jason as part of the crew on the *Argo*. They are also the children of Zeus and Leda, conceived while Zeus was disguised as a swan. In this disguise he is sometimes associated with the constellation Cygnus, bringing yet another constellation we've met already into this story. The twins are often said to have been hatched out of eggs. Another story about the twins draws on the position of the constellation Gemini on the Milky Way. One early view of the Milky Way was that it represented a herd of cattle. The twins' position, half on, half off the Milky Way is said to represent them in the process of running off with the cattle they have stolen.

The depiction of the twins in Bevis's catalogue doesn't play too heavily on the stolen cattle story; certainly the twins do not appear here as though they are running away. Other depictions reveal them in slightly different positions. Here the twins sit side by side with Pollux wrapping his arm around his brother's waist. With his other hand Pollux holds a sickle above his head while Castor holds a stringed instrument in his right hand and a two-headed arrow in his left. While today the twins are universally identified as Castor and Pollux, historically they have also been identified as the gods Apollo and Hercules. Apollo, Artemis' brother and, among other things, the god of archery and music, is often depicted as Castor is here – holding a lyre and an arrow. Hercules used a sickle to attack the

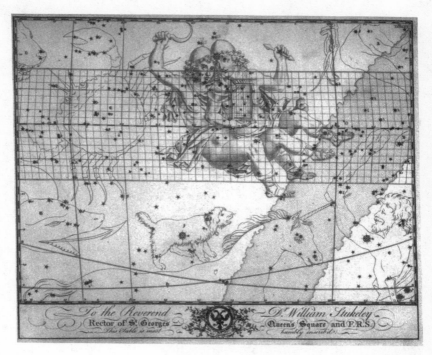

The constellation Gemini

Lernaean Hydra in his second task and so is sometimes depicted with this tool. This may explain the objects Castor and Pollux are holding.

Both Castor and Pollux are recognized by many different cultures. This is partly due to their brightness, but mostly because of their position on the ecliptic, the apparent path of the Sun, Moon and stars. Chinese astronomers identified them as Yin (Castor) and Yang (Pollux) while Vedic (very early Indian) astronomers saw them jointly as one of their lunar mansions, specifically, as Nakshatra Punarvasu.

We have already come across the idea of lunar mansions or lunar lodges in terms of Chinese astronomy and astrology. The Hindu versions of these are different, though, like the Chinese, refer to the changing position of the Moon over a lunar month. The Vedic texts talk about all kinds of things including astronomy and astrology (the two being intricately linked in all cultures until very recently). Our zodiac constellations represent the part of the sky the Sun appears to move through over the course of the year. In Vedic astrology this zodiac is also used, but in addition to these describes twenty-seven lunar mansions. These represent the part of the sky the Moon appears to move through over the course of a month. In Vedic astrology each of those different parts of the sky is a lunar mansion – or nakshatra – marked by a star or group of stars. The first nakshatra is Asvini, the head of Aries.

Each nakshatra has a Hindu myth associated with it. For example, the first nakshatra is Asvini, the wife (or possibly mother) of the Ashvins – twins with the power to usher in the dawn, to turn darkness into light, ignorance into knowledge. They are also associated with medicine and healing. Those born with the Moon in Asvini are thought to be quick and bright and with healing hands. Punarvasu (made up of the stars Castor and Pollux) is the seventh nakshatra, the lunar mansion under which the Hindu god Rama (one of the incarnations of Vishnu) is said to have been born. Punarvasu is sometimes called the star of renewal since it is associated with restoration and the return of good things.

Argo Navis as Noah's Ark

The final constellation connected with Argo Navis and the stories surrounding this ship is Columba, the dove. Johann Bayer illustrates this constellation with a dove holding an olive branch in his beak – in other words Noah's dove – in his 1603 Uranographia, but does not describe it as a separate constellation. Instead, he regards it as part of the neighbouring constellation, Canis Major. The person credited with making Columba or, to give it its full name, Columba Noae, a separate constellation is the French astronomer Augustin Royer.

Bayer, using information from his various explorers and merchants, had classified both Columba and Crux (the Southern Cross) as mere 'asterisms'. In both cases, Royer raised their status to constellation in his 1679 catalogue. Columba can be found on the Bevis plate for Argo Navis, holding his olive branch down in the bottom right-hand corner. An alternative story for Columba, linking it far more heavily with the Argo Navis story, is that this was the bird sent by Jason on the advice of King Phineus through the Clashing Rocks. Given Bevis's depiction of Argo Navis, this seems to have been the story he, at least, favoured.

Not a great deal else is known about Royer, except that he also tried (and failed) to introduce two new constellations in honour of the Sun King, Louis XIV. Already we've met a number of these attempts by astronomers to ingratiate themselves with their patrons or potential patrons by naming something in the sky in their honour. Galileo is perhaps the most famous example, having named the moons of Jupiter the Medici planets in honour of his patrons. This name has not survived the test of time, however, primarily because it was soon discovered Jupiter had many more than just four moons.

The Heveliuses, as we saw in Chapter 1, created the constellation Scutum Sobiescianum (Sobieski's shield) in honour of their patron, the Polish king. Similarly, a few years later, in 1725, Halley named a star in one of Hevelius's constellations, Canes Venatici, Cor Carolis meaning 'the heart of Charles', after Charles I.

But by the end of the eighteenth century the world of astronomy was tiring of this practice. When Augustin Royer tried to create Lilium, the lily, after the emblem of France, and Sceptrum et Manus Lustitiae, the sceptre and hand of Justice, in honour of Louis XIV, they appeared in his own catalogue and nowhere else. Similarly, when William Herschel tried to name his newly discovered planet Georgium Sidus there was only limited acceptance before the name was discarded entirely. Johanne Bode also tried and failed in his effort to create Honores Frederici in honour of the recently deceased King Frederick of Prussia. Astronomy had, it would appear, moved on. The French and American revolutions, supported by many of the men and women of science across Europe, cannot have helped. Naming any kind of celestial body after a monarch or patron was no longer an acceptable thing to do.

Telescopes

So far we've looked only at stars you can see with the naked eye. If you wanted to look at some of these stars, or indeed the Moon or the planets, in more detail you would need binoculars or a telescope. Most astronomers will advise you to go for binoculars and a stand since, for the magnification you get, these are often much cheaper and easier to handle than a telescope. If your heart is set on a telescope, there is a mind- boggling range of options out there for you to choose from.

I don't own my own telescope; the constellations and their history fascinate me more than looking at blurred images of things I would have to know beforehand as a particular type of star. It should be remembered that what you see will be hugely dependent on the magnifying power of your telescope, which in turn will be largely limited by the price you are prepared to pay. If I were to get my own telescope though, I don't think I would go with a Meade or a Celestron or any of the other well-known and well-respected makes of modern amateur telescope. Instead, I would get myself a Porter Garden Telescope – or at least I would if I could afford it.

The Porter Garden Telescope was the creation of explorer, architect and artist Russell Porter. In many ways it looks more like a garden sundial or even some entirely unscientific ornament than a telescope. In a sense, this was Porter's aim: he wanted to create a telescope that could stay out in the garden all year round and that was both beautiful and functional.

Porter joined Frederick Cook's expedition to Greenland in 1893 as the expedition's artist and surveyor and took part in the (unsuccessful) Baldwin–Ziegler 1901 expedition to the North Pole. Back home in Springfield, Vermont, he took up astronomy and telescope-making. In 1919 he was invited to work for James Hartness, a fellow telescope enthusiast but also a successful businessman, as an optical and mechanical engineer. Three years later he formed with some of his students – and with the support of Hartness – the Springfield Telescope Makers Club, later renamed the Stellafane Society, now the largest amateur astronomy society in the world. It holds an annual star party called the Stellafane Convention in August, around the time of the Perseid meteor shower.

Porter began working on designs for his garden telescope in the 1920s. In aiming to create something both beautiful and useful, he was following the philosophy of his contemporary William Morris and those in the Arts and Crafts movement. Like them, he failed or perhaps simply chose not to recognize the importance of afford-ability. One of his garden telescopes costs as much as a relatively expensive new car, as in fact do the accurate and detailed replicas produced today by the company Telescopes of Vermont. However, Porter did succeed in creating something both beautiful and useful, a telescope that needs no setting up or dismantling, that can sit, looking beautiful, in the garden all year round, coming into play whenever there is something in the sky you might want to see in a little more detail. In all he and the Springfield Telescope Makers produced about fifty instruments, of which several now survive in museums, science centres and private collections. Porter went on to design at the personal request of George Ellery Hale a larger, more traditional telescope for professionals, the 200-inch Hale telescope at the Mount Palomar Observatory in California.

The Porter Garden Telescope is hardly representative of ama-teur telescopes either now or then but it does, despite its quirkiness, play a role here, too. Throughout the seventeenth and eighteenth centuries telescopes were made to be beautiful as much as functional and rich amateurs could easily boast a telescope that might rival or even surpass the best owned by the few state-run observatories that then existed. Towards the end of the nineteenth century, however, all that changed. Professional observatories started to commission very large and expensive telescopes quite beyond the reach of even the richest individual amateur. Things got even worse for the amateurs in the late twentieth century as countries banded together to build enormous telescopes on remote hilltops in areas with reli-ably good weather and clear skies. At the same time, the smaller, more affordable amateur telescopes ceased to be beautiful objects

The Porter Garden Telescope

made of brass and glass and became functional boys' toys, affectionately referred to by their owners as 'bits of kit'.

Attempts were made towards the end of the nineteenth century and beginning of the twentieth century to draw in new audiences to telescopes by getting away from traditional design. A comparison could be made with modern technology today and the success of the Nintendo Wii or Apple iPod. In both cases new audiences

FEBRUARY

were pulled in to buy essentially old products by tinkering with design and marketing. The same was true for a small number of telescopes in the late nineteenth and early twentieth centuries, though with admittedly less success. Porter's Garden Telescope is just one example; another was the instrument invented by the telegraph engineer Josiah Latimer Clark.

Josiah Latimer Clark's transit instrument – a telescope pivoted so that it can only go up and down and not side to side – was designed primarily as a means of allowing people to check their watches by the stars. His hope was that 'if this charming instrument were more fully known it would become as popular as the stereoscope or camera'. In the pamphlet accompanying the patent he continued with his assertion that 'many a lover of nature will find immense pleasure in watching the exquisite precision of the movements of the heavenly bodies'. His aim in the pricing and marketing of his telescope was to make it so commonplace that people would hardly see it as a scientific instrument. In his pamphlet he makes casual comparisons with all kinds of domestic gadgets, his hope being that it might appeal to those 'who would never dream of opening a book on Astronomy but who may desire to use a transit instrument as they would use a telephone, without necessarily caring to study the principles of its construction'. Sadly, like Porter's Garden Telescope, it never really took off.

But, for the public at large, telescopes have a fascination, both as modern pieces of precision engineering linking us with the stars and, increasingly as they were in the eighteenth century, as objects of beauty and erudition.

CHAPTER TWELVE

March, Astrology and the Zodiac

THE FAMOUS EIGHTEENTH-CENTURY nonconformist English
hymn-writer Isaac Watts wrote a poem to help children
remember the official signs of the zodiac:

> The Ram, the Bull, the Heavenly Twins,
> And next the Crab, the Lion shines,
> The Virgin and the Scales;
> The Scorpion, Archer, and the Goat,
> The Man that pours the Water out,
> The Fish with glittering scales.

ISAAC WATTS (1674–1748)

Isaac Watts wrote many hymns that are still sung today, but is probably best remembered for his poem 'Against Idleness and Mischief',
parodied in *Alice in Wonderland* as 'How doth the Little Crocodile . . .?'
None of his work was concerned exclusively or even primarily with

astronomy or astrology, but rather with education. Like the many poems in his books, this one was included because Watt thought knowledge of the zodiac was something every child should learn.

At any time of year, from any location, some of the ecliptic and so a few of the zodiac constellations are generally visible. In March, for instance, you might be able to see Aries, Taurus, Gemini, Cancer, Leo and Virgo. These constellations are some of the best known in the night sky – not familiar as shapes like Ursa Major and Orion, but as familiar astrological names. They are, and always have been, central to astrology because they mark out the ecliptic, the band of sky through which the Sun, Moon and planets appear to wander. They mark out the part of the sky where things are most likely to happen, and so where messages for astrologers to interpret are most likely to appear.

Whether you believe in astrology or not, most people in the Western world know their star sign. Mine is Scorpio, though even in my teens (an age when girls seem especially susceptible to believing in horoscopes) my ability to believe was slightly tempered by the fact that my dad and granddad were also Scorpios. Whatever was predicted for me must also be true for both of them; and this thought rather destroyed the illusion for me.

Today's astronomers are armed with defences against astrology – why it can't work and so should never be confused with astronomy. The first and most obvious objection challenges the idea peddled by newspaper and magazine horoscopes that one-twelfth of the population should have the same personality and destiny. But of course astrologers counter this quite easily by making a distinction between those horoscopes created to fulfil the commercial needs of the publishing industry and 'real' horoscopes that are tailored to the individual. The other main arguments against astrology are a little harder to counter and tend to stress the importance of new discoveries made since the rules of astrology were set down.

In the first place, astronomers question how astrology might actually work. If the planets have some influence on our lives they must, they reason, exert some force upon us, but what force could that possibly be? There are only two known forces that can act over the kinds of distances that separate us from the planets, gravity and electromagnetism, and both get weaker with distance. In terms of gravity, the pull of the planets is minuscule compared to that of the Moon, since the Moon, though not as massive, is so much closer. Similarly with electromagnetism, the effects of the Sun's magnetic field, which, as we have seen in Chapter 3, does sometimes cause things to happen on Earth, should drown out any effects caused by the planets. The Sun is much larger than the planets and has a much stronger magnetic field. It is also closer to us than many of the planets.

In astrology, however, all planets have the potential to affect things on Earth in equal measure – their distance is irrelevant – which suggests they must be acting through some unknown force. If this is the case, astronomers continue, what about all the newly discovered planets orbiting other stars or even the thousands of asteroids (essentially small planets) we find in our own solar system?

Finally, astronomers will point to Ophiuchus, the thirteenth constellation of the zodiac. Precession, the wobble of the Earth, has caused the path of the Sun, Moon and planets to gradually shift in relation to their background of stars over hundreds of years. This means that Ophiuchus is now one of the constellations that these bodies appear, from Earth, to pass through. This means that birth dates and the zodiac constellations astrologers attach to them no longer match up with the position of the Sun. Ophiuchus' new status as the thirteenth constellation of the zodiac was only made official with the IAU's boundary definitions for each constellation, though it was known as far back as the second century CE when Ptolemy charted Ophiuchus as crossing the ecliptic. In his treatise on astrology (Tetrabiblos), however, Ptolemy doesn't count it as a zodiac constellation; and that has been the tradition in astrology ever since.

Modern astrologers often defend themselves by simply pointing to their success rate. Why should we get bogged down in how it works, they argue, when there are so many things in this world science cannot yet explain, why not just accept that it works? And that of course is the essential point: if you believe it works then any argument about the technical impossibility of it working can be easily dismissed as closed-mindedness. If you don't believe it works, the technical arguments can be very persuasive.

Astrology has a very long history, and was for a long time seen as intricately linked to astronomy. Earlier astrologers, who were often also astronomers, were more concerned with explaining how astrology worked than modern astrologers. The first astronomical records were kept to help astrologers in their work, which was primarily to read and understand omens. The Mesopotamians, and more specifically, the Babylonians, saw the heavens and indeed all nature as a kind of message board of the gods; this was their explanation for how astrology worked. The gods caused everything that happened on Earth – floods, famine, the birth or death of a king – and it was up to a particular class of expert called ummanu to read the messages the gods sent and recommend action that would avert any impending disaster. For example, eclipses were regarded as particularly powerful signs and dangerous to monarchs, so much so that it was standard practice for the monarch to go into hiding during an eclipse while a dummy-king sat on the throne.

The word 'planet' comes from the Greek meaning 'wanderer'. The Sun, the Moon and the five planets you can see with the naked eye – Mercury, Venus, Mars, Jupiter and Saturn – all appear to move independently against a backdrop of 'fixed' stars, while the stars all appear to move together. The stars have always been referred to as fixed, in contrast to the wanderers that we now know to be bodies within our solar system. The stars were fixed because they were unchanging and always stayed the same distance from one another. This independent movement of the wanderers made them the assumed carriers of most of the gods' messages. It was noticed early on that, although the planets moved independently, they were all confined to a particular band of sky, the ecliptic. The Mesopotamians divided the ecliptic gradually, over thousands of years, into eleven animals and people (Libra, the scales, was added later by the Greeks), our current zodiac. Other cultures have divided them differently. The lunar mansions from Hindu or Vedic astronomy that we met in the last chapter divided the sky into twenty-eight constellations.

Bevis's illustration of the sky north of the ecliptic (see overleaf) shows the zodiac constellations cut in half, marking out the boundary of the circular star chart. This map, based on Ptolemy's as Bevis states in the top left-hand corner, shows Ophiuchus just on the ecliptic between Scorpius and Sagittarius, while Libra, the scales, is not there at all.

The zodiac constellations

Though all but one of the twelve constellations originated in Mesopotamia, they did not arrive all at once. A popular theory on how the constellations came about, as was mentioned briefly in

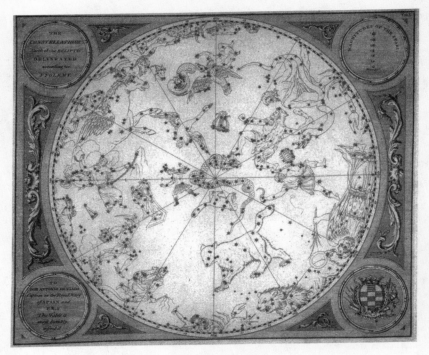

The sky north of the ecliptic

Chapter 7, groups them into three groups of four created around 3,000 years apart. In each group there is a constellation that is symbolically connected with spring, another with summer, one with autumn and the last with winter. Gemini, Taurus and Aries, for example, are all associated with spring. Similarly, the Bull and Ram are frequently associated with male fertility and so the very beginning of life. Twins, such as the twins in Gemini, feature in the creation myths of many cultures – the separation of the earth from the sky, day from night, land from sea and so on. Virgo, Leo and Cancer are associated with the summer. Virgo, generally depicted with an ear of grain, represents a sort of Mother Earth, summer fertility figure. Lions are generally associated with power and importance, as reflected in the Sun reaching its highest point in the sky in midsummer. Cancer also represents the highest point of the Sun, but its focus is on its change of direction – getting higher day by day and then lower – rather than the absolute height.

Sagittarius, Scorpius and Libra are our three autumnal constellations. Sagittarius and Scorpius both symbolize the beginning of a downfall, the shot of an arrow, the sting of a tail causing the Sun to begin its downward trajectory. Libra, meanwhile, can be read as symbolizing the balance between day and night at the autumnal equinox. Pisces, Aquarius and Capricorn are our winter constellations. In each case, water plays a key role, as it also does in the iconography associated with the underworld, into which the Sun was thought to descend.

These groups, so the theory goes, all correspond to a different point in history where they marked the different astronomical points – the solstices and the equinoxes. But, as we've seen, these shift due to precession; and so, every so often, a new set of constellations would be needed to mark the same astronomical points. By this theory the creation of the different constellations can be dated to around 6500 BCE for Gemini, Virgo, Sagittarius and Pisces; 3500 BCE for Taurus, Leo, Scorpius and Aquarius and around 1000 BCE for Aries, Cancer, Libra and Capricorn. A quick glance at these dates will tell you we should be due for a new set around now, and in a way we are except that rather than create a new set we simply move round one. Pisces now houses the Sun at the spring equinox, Gemini marks the summer solstice and so on. As well as being a very neat theory, it does actually match up to what we know about the origins of the constellations. Libra for example, the only zodiac sign to be of an inanimate object, never appeared in star charts or tables before the date at which the Sun is in Libra at the autumnal equinox in the northern hemisphere.

Astrology's past

In early civilizations the sky was thought only to contain messages about big events and the lives and fortunes of the monarchs. Later this was extended to the health of everyone, or at least everyone who could afford a doctor, and later still to the lives of anyone who could afford the services of an astrologer. When astronomers did eventually start to distance themselves from astrology and publicly criticize it, the way of thinking was too ingrained to change: by then most people had some sense that the stars and planets affected their lives, and this has been surprisingly hard to shift.

When the Greeks took over Mesopotamian astrology they adopted the constellations and the idea that each planet (including the Sun and Moon) represented a different god but then integrated some of their own ideas into the mix. Already the Greeks had an understanding of geometric relationships (working out alignments and such like), and a theory about the basic building blocks of all things, namely earth, fire, air and water. These theories were incorporated into Greek astrology, so that when Marcus Manilius, writing not long before Ptolemy, set out the rules for astrology they were quite lengthy. Annoyingly, he gave no reasons for how the new rules worked.

Ptolemy did attempt to give some rationale for astrology in his treatise Tetrabiblos. The Moon affects tides, the Sun affects the seasons, climate and the growth of plants; was it not then logical to assume they and the other heavenly bodies affected the personalities, health and fate of human beings? This did not call for such a stretch of the imagination as all that. Medicine and astrology were already closely linked by the time Ptolemy was writing. Hippocrates, working as a doctor in around 460 BCE, was greatly influenced by the astrological ideas coming to Greece from Mesopotamia. Today he is best known for instituting the Hippocratic Oath. But he is also credited with the creation of an astrological system of medicine that persisted in Europe until well into the sixteenth century.

According to Hippocrates, the planets alter the balance of the four humours (blood, black bile, yellow bile and phlegm) in the body. The humours were associated with the four elements (earth, fire, water and air) which in turn were associated with different planets. The zodiac was seen as governing different parts of the body – starting at the beginning of the zodiac with Aries and the head, ending with Pisces at the toes – while the planets were each dominated by a particular element or elements. The position of the planets in a particular sign of the zodiac would then strengthen or weaken the power of the elements associated with that planet on the part of the body associated with that sign, and so affect the balance of the humours in that body part. This could be treated or counterbalanced by the use of a plant or mineral, which were also associated with particular signs or planets. Planets, Ptolemy argued, radiated their influence as the Sun radiates light and heat. How you were affected by these particular alignments depended on the arrangement of the stars and planets at the precise moment of your birth or conception.

This idea about the body being governed by different zodiac constellations that Hippocrates set out in the fifth century BCE was still going strong more than two thousand years later. More than anything else, it was this link with medical theory that gave astrology its power – it could be used to understand and treat disease. This allowed it to survive in medieval Europe even when Christianity essentially banned all the future-predicting elements of astrology. In the Islamic world, where the future-predicting elements of astrology were also frowned upon for religious reasons, astrology was given a new dimension. A number of important astronomer/astrologers began working out links between historical events and astrological alignments – like today's astronomers who date early manuscripts by cross-checking them against known astronomical events mentioned in the text. They also worked on the details of how to read the messages in the sky, and this was transferred back to Europe.

According to Hugo of St Victor, writing in the twelfth century, astronomy was simply the practical arm of astrology – astronomers collected the data, astrologers interpreted them. By the fifteenth century astrology had become institutionalized in Europe, in that it was taught in universities and every court had its own astrol-oger. With the arrival of printing, from about around 1460 onwards predictions began to be printed in papers and pamphlets (later to take the form of popular almanacs) covering everything from the weather to the next natural disaster or visitation of the plague.

During this period, explanations of how astrology worked took a slightly odd turn. A set of manuscripts was discovered, referred to as the Hermetic texts, purporting to pre-date Moses. Because of the Renaissance belief that the earlier the work the more reliable it was likely to be, these manuscripts became a highly regarded source of information about astrology. The texts were later discovered to be from the second century CE – after Ptolemy – but by then the ideas they contained were in wide circulation. They describe how the heavens were thought to interact with the Earth and so how astrology worked. The key feature of this Hermetic universe was the presence of spirits or demons that are everywhere and carry influences with them. The planets and the signs of the zodiac all radiate these demons, while different astrological combinations produce different types. However, the demons did not entirely rule the lives of men and women on Earth since, with proper training of the intellect, you could rise above these influences, just as you could rise above pure instinct.

In the sixteenth century the audience for astrology expanded. Not only were there astrologers in all the major courts and universities of Europe and the Middle East but still more were setting up private practices as astrologer-physicians. Shakespeare makes numerous references to astrology in his plays, assuming a familiarity with those terms from his audience.

With this new popularity came an increase in polemics against astrology, especially in Europe. Protestantism challenged the authority of the Catholic Church, which became less secure in its position. Astrology began to be linked with magic and alchemy in treason cases. Predicting a monarch's downfall was considered to be treasonous. Some believed that the popularization of astrology – in the form of pamphlets and almanacs – had robbed it of its intellectual integrity and opened it up to these kinds of suspicions. Certainly there was money to be made as an astrologer, and no restrictions on who set themselves up as one; but, having said that, the same was true of doctors, who suffered no such criticisms.

A distinction was made between jobbing astrologers, who were considered to be charlatans, and respectable astronomers. Johann Kepler, Tycho Brahe's assistant and successor at the court of

Emperor Rudolf II, was one such figure. Today he is best known for his laws defining how the planets move around the Sun and the shape of their orbits, all pre-dating Newton, but in his own day he was just as well known for his horoscopes. While he was dismissive of many jobbing astrologers, those who he felt did not fully understand the scientific basis of their discipline, he was a supporter of astrology, supplementing his own income by casting horoscopes. He once described astronomy as the wise but poor mother who would starve should her daughter, astrology, not support her. Some four hundred of his horoscopes are thought to have survived in libraries around the world.

Brahe, Kepler's boss, also accepted the principle that the stars influenced life on Earth but disputed the claim that their influence was absolute. He was sceptical about some of the claims of contemporary astrologers and about some of the interpretive systems used to read the stars, but he was not against astrology per se. When he discovered a supernova in 1572 he produced astrological interpretations of it, as he did for the comet of 1577.

Then, quite suddenly, between around 1650 and 1700, astrology lost its scientific respectability. There is no evidence to suggest a single pivotal moment — astronomers did not suddenly start denouncing astrology, though they did occasionally make fun of it in private. John Flamsteed, for example, Britain's first Astronomer Royal, cast for the entertainment of his friends a now relatively famous spoof horoscope of the Royal Observatory, Greenwich on its establishment (or birth) in 1675.

Nor was there any single astronomical development in that period that once and for all disproved the basis of astrology. Rather, the split between astrology and astronomy appears to have evolved over a couple of generations as a result of a number of broadly simultaneous developments. For one thing, a Sun-centred universe had by 1650 become the standard model, which changed the nature of what the constellations, including the zodiac, signified. The constellations were no longer absolute arrangements but depended on our own perspective. The invention of the telescope and its application in astronomy showed there were features to the heavenly bodies never previously considered by the astrologer. Comets, too, lost some of their power as warnings of impending disaster when Isaac Newton and Edmund Halley discovered that their paths, like those of the planets, were regular and could be predicted. Newton's laws of gravity also took some of the power away from

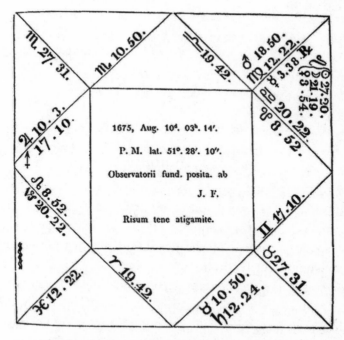

John Flamsteed's comic horoscope of the Royal Observatory

astrological forces (though, interestingly, many of the arguments against gravity at the time claimed that it involved force at a distance without any visible means of transporting that force). Finally, and perhaps most importantly, there was a change in attitude as to how science should be performed – in effect, what science was. This new science was (at least rhetorically) about collecting quantitative evidence, about being able to measure how one thing affected another. Astrology, almost by definition, does not and never can deal in quantitative, measurable absolutes.

Until this point of separation astronomy owed much to astrology. For hundreds of years astronomers and astrologers studied the same part of the sky, used the same tables and were even often the same people. Today, even though there are many more scientific arguments against astrology, often the debates are couched in just the same terms as they were five hundred or even a thousand years ago. Can you and should you predict the future? And, for all the arguments against astrology, the practice has never completely gone away.

Mythology

We have met already in earlier chapters most of the thirteen zodiac constellations and their associated mythology. Aries, the bull, is commonly associated with the mythical ram whose fleece became the Golden Fleece. Taurus is often associated with the Minotaur slain by Theseus in the story leading up to Ariadne gaining her crown, Corona Borealis. He is also sometimes associated with the Cretan bull captured by Hercules in the seventh of his twelve tasks. Either way, Taurus is generally depicted, as Bevis depicts him opposite, as half a bull with his back legs missing – which perhaps

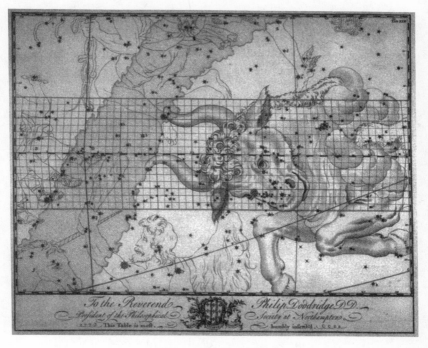

The constellation Taurus

adds credence to the Minotaur association, since the Minotaur was half-bull, half-man.

Gemini, the twins, we know as Pollux and Castor, two of Jason's Argonauts. We met Cancer in Chapter 2 as the heroic crab sent by Hera to attack Hercules only to get crushed beneath his feet without his even noticing. Leo also has a place in the Hercules story, as the Nemean lion Hercules is sent to kill and whose skin he is then depicted wearing. Virgo, meanwhile, is thought to be associated with Persephone, who rises from the underworld in the spring.

Libra was invented only in Roman times and so does not have any Greek mythology associated with it, or any Roman stories either. The stars making up Libra formed part of Scorpius' claws in Ptolemy's star chart. The scorpion, Scorpius, does feature in Greek mythology – again in a story we have already heard – as the successful assassin of Orion. Sagittarius is associated with various centaurs in Greek mythology but most frequently with Crotus (the archer).

Crotus was the son of the goat-god, Pan, god of shepherds and music, and the nymph Euphemes. Since Euphemes was nurse to the nine muses of arts and science, Zeus and Mnemosyne's nine daughters, Crotus was brought up with them. The muses included Urania, muse of astronomy and astrology, and Melpomene, muse of tragedy. Melpomene is today the name of an asteroid discovered by J. R. Hind at the Royal Observatory, Greenwich, in 1852. The then Astronomer Royal George Biddell Airy was asked to name it. His daughter had just died and so, in her memory, he named the asteroid after the muse of tragedy. The muses played a part in placing Crotus among the stars. Petitioning their father, they persuaded him to turn Crotus into a constellation on his death and so he became Sagittarius. Some say the smaller constellation, Sagitta, is an arrow shot from his bow.

Capricorn is Crotus' father, Pan. Running from Typhon, the god of storms, a monster and enemy of the Olympian gods, Pan, who on land was half-man (the top half), half-goat, dived into the Nile. On hitting the water his top half became goat, and the rest fish. It is in this form, half-goat, half-fish, that he was placed in the sky as the constellation Capricorn. Aquarius is generally associated with Ganymede, the wine-bearer of the gods whom we met in Chapter 2 in relation to the eagle Aquila who brought him to Mount Olympus. Finally we have Pisces. Here the two fish represent Eros, god of love, and Aphrodite, goddess of lust and beauty, who dived into the water to escape, like Pan, the storm god Typhon, only to be turned into a fish.

Evolving stars and pulsars

To modern-day astronomers the zodiac constellations are just like any other constellation, made up of stars that come in various different types. Some are nebulae, supergiants, star clusters and, in Taurus, one of the first pulsars to be discovered. Taurus is in fact a great place to start looking at the stars within the zodiac constellations since it has a great deal to offer the stargazer.

Taurus is a very bright constellation, best seen from the northern hemisphere in the winter. Its brightest star, Aldebaran, is a red giant and the thirteenth brightest star in the whole night sky. Making a 'V' shape with Aldebaran are some very bright stars as well as some fainter ones that make up an open star cluster known as the Hyades. Because they are so easy to see with the naked eye the Hyades have been known since antiquity and, like the full constellations, have Greek mythology associated with them. According to myth, the Hyades are the daughters of Atlas and half-sisters to the Pleiades (another star cluster found in Taurus). When the Hyades' brother Hyas died they were turned into this constellation. Their appearance is associated with rain, representing the tears they shed for their dead brother.

The Pleiades, the Hyades' half-sisters, are another open cluster found in Taurus and are sometimes called the seven sisters. You can find them near the stars that make up the general shape of Taurus rather than as part of that shape, and they are often marked as a separate constellation on early maps. You can see the Pleiades marked on Bevis's atlas, in this instance as part of Taurus, towards the top of his neck, near the cloudy part marking the end of this half-bull.

As you'll remember, open clusters like the Hyades and Pleiades are groups of a few thousand young stars all originally formed together out of the same cloud of matter. This means they're all roughly the same age, mostly they're 'young' stars, by which astronomers mean less than a few hundred million years old. The Pleiades are one of the best-known open clusters in the night sky. They are nearer than most and are easily made out with the naked eye as a distinct group. Most cultures around the world seem to

The Pleiades open star cluster, photographed by amateur astronomer and photographer Robert Gendler

have a story connected to them, and most of the world's major religions, too. They are mentioned several times in the Bible; they are associated with the Hindu war god, Skanda, as his six mothers, while some scholars of Islam have associated them with al-Najm, the star mentioned in the Qur'an.

Besides these two open clusters, Taurus is also home to the Crab nebula. Though not visible to the naked eye it is worth a mention for its importance in the history of astronomy. The Crab nebula is a supernova remnant. It is essentially the outer layer cast off by a very massive star when that star became a supernova and is made up of all the different elements created in that star – helium, carbon, oxygen and so on – as well as any hydrogen the star didn't use. Out of this cloud of cast-off material new stars and even new solar systems now begin to form.

The star that created the crab nebula became a supernova in 1054. The supernova was recorded by Chinese astronomers, though astronomers in Europe missed it entirely despite its being visible during the day for a period of around three weeks. The Europeans missed it for the same reason they nearly missed Tycho Brahe's supernova in 1572 – their view of the universe did not allow for any changes among the fixed stars, which were unchanging and perfect. A sudden, dramatic change in the brightness of a star did not fit with that view; so they didn't look for it and didn't see it.

John Bevis was the first European to record this nebula, in 1731. A century later it gained its name, the Crab nebula, when the Irish astronomer William Parsons, Earl of Rosse, drew what he saw through his giant telescope at Birr Castle and thought it looked like a crab. Rosse's giant telescope was for a time the largest in the world and is currently being restored at Birr Castle for visitors to marvel at, as they did in its heyday in the 1840s. In the early twentieth century the nebula became one of the first to have its original supernova identified (and so provide some indication of how some nebulae are formed) thanks to the records kept by Chinese astronomers.

In the middle of the Crab nebula are the remains of the core of the star that became the supernova which then created the nebula. This core is one of the first and most famous pulsars to be discovered. The very first was PSR 1919+21 in Vulpecula, discovered by Jocelyn Bell, a PhD student, and her supervisor, Antony Hewish, in 1967. They originally called it LGM-1, short for 'little green men – 1', after one of the possible (though unlikely) sources for the signal they were receiving.

Pulsars are radio sources that give out a very regular pulse. They are thought to be very fast-spinning neutron stars with strong magnetic fields. As they whiz around they send out a beam of radio waves, rather like a lighthouse sending out a beam of light. A neutron star, first discovered in 1933, is a very dense core of a star left over when a very massive star becomes a supernova. Even more massive stars will become black holes. The core in a neutron star is so dense that the atoms within it break down into their constituent neutrons, protons and electrons. The protons and electrons then combine to become neutrons until the whole core is made up of nothing but neutrons – hence the name.

As suggested by their half-joking name 'little green men – 1', when Jocelyn Bell and Antony Hewish discovered their pulsar they entertained the possibility, alongside many other options, that the pulses might be a message from life elsewhere in the universe. They happened to mention this in their paper in the journal *Nature* and immediately attracted the attention of the press. Once the press discovered a young, attractive woman was behind the discovery they made it a photo shoot as well and Bell was required to pose poring over various mocked-up charts looking intellectual. In 1974 Antony Hewish, but not Jocelyn Bell, was awarded the Nobel Prize for Physics for the discovery. Various people have argued this was unfair; Jocelyn Bell herself has not. She has pointed out how unusual it would be for a research student (regardless of their sex) to be given the Nobel Prize and that the supervisor is and should be the one with final responsibility – for blame or credit – over a student's project.

The pulsar in Taurus, in the centre of the Crab nebula, was discovered just one year after Bell and Hewish's discovery. Through it, the possible connection between pulsars and supernovas was made. If a pulsar was found at the centre of a nebula that had definitely resulted from a supernova, it seemed likely the pulsar originated from that supernova, too. This has indeed been found to be the case.

After Taurus, the stars visible in the other zodiac constellations can seem something of an anti-climax. The stars in the relatively bright constellations Aries and Gemini we met in the story of Jason and the Argonauts. Cancer, as we saw in Chapter 2, contains the open cluster M44, otherwise known as the Beehive cluster.

Leo is the home of the Leonid meteor shower that comes around every November and of the very bright star Regulus. Virgo is a very large constellation stretching a long way round the ecliptic with some relatively bright stars. The brightest by far is Spica, a blue giant and also Cepheid variable. In the Bevis depiction overleaf, as in most illustrations of Virgo, Spica marks the ear of corn Virgo is holding in her left hand.

Spica's main claim to fame is as the star the Greek astronomer Hipparchus is thought to have used to discover the precession of the

MARCH

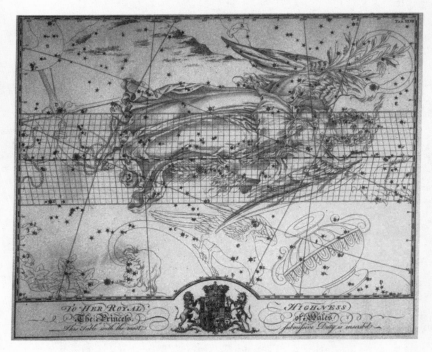

The constellation Virgo

equinoxes. Precession causes our alignment with the ecliptic to change over time and so for the solstices and equinoxes also to shift. This is why new constellations needed creating about every three thousand years to mark these points and so give us the twelve zodiac constellations. Until the time of Hipparchus, however, no one knew that this was happening; it was simply thought that at different stages the existing constellations marking the solstices and equinoxes were a bit inaccurate and that new ones were needed. Writing around 130 BCE, Hipparchus compared the positions he measured for Spica and other bright stars with those measured by previous astronomers and worked out that the equinoxes were moving through the zodiac by not less than 1° each century. That's 1° of a total 360° around the whole circle of the zodiac. Today the figures indicate a shift of around 1° every seventy-two years – or one whole zodiac constellation approximately every 2,160 years.

The small and quite faint constellation Libra follows Virgo around the zodiac and is then followed by Scorpius, which has some bright stars including the very bright Antares. Antares is a super-giant with a companion star, making it a binary or double star. In Ophiuchus is the Barnard star, best known as the first place Arthur Dent and Ford Prefect attempt to hitch a lift to in *The Hitchhiker's Guide to the Galaxy*. To astronomers, at least those who have managed to avoid *Hitchhiker* entirely, it is perhaps better known as the place in which the first planet outside our solar system was nearly found. In 1990, observers of the Barnard star noticed a wobble that might possibly have been the result of a planet orbiting the star. Later observations found the research to be flawed.

Another star of note in Ophiuchus is a recurrent nova, RS Ophiuchi. This star changed suddenly from magnitude 12 to something visible to the naked eye and then back again. It does this every twenty years. Its next sudden brightening is due in 2025. Sagittarius follows Ophiuchus and corresponds, as we saw in Chapter 6, with the middle of the Milky Way. It is also the home of lots of nebulae, but none visible to the naked eye. Sagittarius is followed by three really rather faint constellations, Capricorn, Aquarius and Pisces. Though no spectacular naked-eye stars stand out among these three, the blue-white supergiant in Aquarius with an apparent magnitude of 2.9 is pleasingly named Sadalsuud, meaning 'luckiest of all'.

The zodiac, different parts of which are visible all year round, provides a neat place to end this stargazer's guide. The zodiac constellations bring together many of the different types of star and several of the Greek myths found in this book. What's more, because of its position as 'home' to the planets, the Sun and the Moon, the zodiac has played a particularly important role in the mythologies of cultures around the world. Our fascination with the night sky goes back millennia and is still going strong today. While the ancients projected images of heroes and monsters on to the stars, we now dream of travelling among them or, at the very least, understanding what they are – and so what we, products of stardust ourselves, are made of.

Afterword

THIS BOOK IS TO AN EXTENT a reflection of my own interests and preoccupations. Though I have aimed to be completist about the constellations, including every one of the eighty-eight now internationally recognized in the night sky, the rest of the book is more idiosyncratic. Inevitably it reflects my own particular fields of study and research – the Herschels, the subject of my (still unfinished) doctoral thesis, and the history of the collection and employees at the Observatory where I worked for many years. My interest in astrology comes partly from a love of women's magazines, but probably also from working for so long among astronomers whose anti-astrological stance made finding its redeeming features almost a sport.

Otherwise, what I hope will come through here are a number of themes that I have come back to again and again throughout this book. Where possible I've tried to draw out the stories of individual people, especially the female stargazers who so often helped their husbands, brothers and sons without getting much credit at the time. Another theme is the relationship between cultural need and the constellations. Hunter-gatherer societies used the stars as a means of signalling which foods would be in abundance and when and where to look for them. Settled farming communities needed the stars to tell them how and when to cultivate crops. Rulers of more complex bureaucratic societies, meanwhile, looked to the stars in an attempt to ensure the survival not just of their food supplies but of their positions within society. Navigation became important as a means of searching safely further afield for supplies; and astrology thrived as a means of justifying the actions of the powerful. As such, a role was created for court astrologers, records were kept and modern astronomy and astrology essentially began. Finally, knowing about astronomy – as distinct from astrology – and taking part in star-gazing became fashionable in the eighteenth century and worthy in the nineteenth. Stargazing in the twentieth century became more a niche activity, though still retaining some of its popular appeal, while space travel meant that astronomy, if not stargazing, was rarely out of the news. Now we're well into the twenty-first century and astronomy is on the rise once more. Following the recent eclipses and the 2004 transit of Venus, the international year of astronomy in 2009 celebrates four hundred years of the telescope and forty years since Neil Armstrong first walked on the Moon. Travel is simpler than ever before, making eclipses easier to chase, dark skies easier to find and ancient observatories easier to visit. From the big cities, too, modern stargazers can get to know our brighter stars and constellations, including of course our Sun, and join in the many events planned worldwide to celebrate 2009. If ever there was a time to get involved in stargazing, this is it.

Travel is another theme that has featured throughout this book. Astronomy provides some fantastic ideas for travel. Whether to see constellations not visible from where you live, an eclipse or some historical site connected with astronomy's long and rich history, there are always plenty of excuses for the stargazer to get packing. And stargazing's potential as a leisure activity doesn't stop there. Though the heyday of collectors might have been in the seventeenth and eighteenth centuries with the collections that have formed the basis for some of our most impressive museums, collecting is still a popular hobby today. There are some beautiful objects to collect – from the tools of the trade such as astrolabes and telescopes to actual pieces of astronomical material such as meteorites.

Finally, I hope this book has left you with some idea of the huge range of objects in the sky that we just casually lump together as 'stars'. Though not exhaustive, I have tried, where examples are visible to the naked eye, to include as many different types of star as possible, from the very young nebulae and open clusters to the very old giants, dwarfs and black holes. Though most of us have come across these terms from time to time, I hope this book has given you a clearer idea of what they actually are and how they fit into the bigger picture – of what you can see in the night sky and exactly how we know what (little) we do about it all.

Of course, the fascinating thing about astronomy is that new discoveries are being made all the time. There's a whole universe out there to explore and understand. Perhaps you'll be one of the next amateur discoverers, or one of your children will be – encouraged by the replica constellation cards you made them when they were five. In the meantime, happy stargazing.

Further Reading

Allan, Richard Hinckley, *Star Names: Their Lore and Meaning* (1899)

Bakich, Michael E., *The Cambridge Guide to the Constellations* (1995)

Bartram, Simon, *Man on the Moon* (2002)

Chapman, Allan, *The Victorian Amateur Astronomer: Independent Astronomical Research in Britain, 1820–1920* (1998)

Curry, Patrick, *Prophecy and Power: Astrology in Early Modern England* (1989)

Dunn, Richard, *The Telescope: A Short History* (2009)

Evans, David S., *Under Capricorn : A History of Southern Hemisphere Astronomy* (1988)

Fara, Patricia, *Pandora's Breeches: Women, Science and Power in the Enlightenment* (2004)

Forbes, Eric C., A. J. Meadows and Derek Howse, *Greenwich Observatory*, 3 vols (1975)

Al-Hassani, Salim T. S. (ed.), *1001 Inventions: Muslim Heritage in Our World* (2006)

Holmes, Richard, *The Romantic Poets and Their Circle* (2005)

Hoskin, Michael (ed.), *Caroline Herschel's Autobiographies* (2003)

Hoskin, Michael, *The Herschel Partnership* (2003)

Hunter, Lynette and Sarah Hutton, *Women, Science and Medicine* (1997)

Impey, O. and A. MacGregor, *The Origins of Museums: The Cabinet of Curiosities in Sixteenth and Seventeenth-century Europe* (1985)

Jeffers, Oliver, *How to Catch a Star* (2004)

Levy, David, *Starry Night: Astronomers and Poets Read the Night Sky* (2001)

Lubbock, Constance, *The Herschel Chronicles* (1933)

Moore, Keith, 'Space and Time Forgot: John Herschel as a Poet', Paper presented at the John Herschel 1972–1871 Bicentennial Commemoration, Royal Society, London, 1992

Nicolls, Helen and Jan Pienkowski, *Meg on the Moon* (1976)

Olson, D. and M., 'William Blake and August's Fiery Meteors', in *Sky and Telescope* (August 1989)

Pattie, T. S., *Astrology* (1980)

Plunket, Emmeline, *Calendars and Constellations of the Ancient World* (1903)

Ridpath, Ian, *Star Tales* (1988)

Saliba, George, *History of Arabic Astronomy: Planetary Theories During the Golden Age of Islam* (1995)

Sutton, Geoffrey, *Science for a Polite Society* (1997)

Tester, Jim, *A History of Western Astrology* (1987)

Uglow, Jenny, *The Lunar Men* (2002)

Walker, Christopher (ed.), *Astronomy Before the Telescope* (1996)

Whitfield, Peter, *Astrology: A History* (2001)

Magazines

Astronomy
Astronomy Now
Sky and Telescope

Useful websites

Aboriginal Astronomy Project:
www.atnf.csiro.au/research/AboriginalAstronomy/about.htm

Astronomy Now: www.astronomynow.com

Fred Espenak's eclipse pages:
eclipse.gsfc.nasa.gov/eclipse.html

Galaxy Zoo: www.galaxyzoo.org

Robert Gendler's astronomical photography:
www.robgendlerastropics.com

International Occultation Timing Association (IOTA):
www.lunar-occultations.com/iota/iotandx.htm

NASA: www.nasa.gov

National Maritime Museum (especially the 'Collections Online' pages):
www.nmm.ac.uk/collections

Royal Astronomical Society: www.ras.org.uk

Sky and Telescope: www.skyandtelescope.com

Picture Credits

Every effort has been made to locate the rights holder to the pictures and figures appearing in this book and to secure permission for usage from such persons. Any queries regarding the usage of such material should be addressed to the author c/o the publisher.

'The Kentish hop merchant and the lecturer on optics'. Illustration. (© *The College of Optometrists*) 3

Composite image of the world at night. Photograph. (*This image comes from NASA's Visible Earth team http://visibleearth.nasa.gov/*) 13

Image of the UK at night. Photograph. (*Reproduced here with kind permission from the Campaign to Protect Rural England [CPRE]*) 14

The constellation Ursa Major. Illustration. (*This image, and all those from the Bevis atlas featured elsewhere in this book, was reproduced with kind permission from the Manchester Astronomical Society and in particular MAS member, photographer and amateur astronomer Michael Oates*) 22

The Sun, the Earth and the seasons. Illustration. (© *Greg Smye-Rumsby*) 26

The constellation Scutum. Illustration. (*Bevis atlas, plate 16, MAS*) 31

The constellation Crux. Illustration. (*Bevis atlas, plate 49, MAS*) 34

The constellation Hercules. Illustration. (*Bevis atlas, plate 7, MAS*) 41

The constellation Draco. Illustration. (*Bevis atlas, plate 3, MAS*) 42

The Hercules globular cluster. (*Photograph from NASA's Hubble Space Telescope*) 48

Hydra with Corvus and Crater on its back, and Sextans near its head. Illustration. (*Bevis atlas, plate 44, MAS*) 51

Astronomers at work with a variety of typical contemporary instruments. Painting. (© *Ann Ronan Picture Library/Heritage-Images*) 57

An astronomical tour map of London. Illustration. (© *Greg Smye-Rumsby*) 60

Astronomers sitting in a box. Photograph. (*Edward Maunder, © National Maritime Museum, Greenwich, London*) 73

A total solar eclipse, photographed in Dundlod, India, by Fred Espenak on 24 October 1995. (*Photograph courtesy of Fred Espenak, NASA/Goddard Space Flight Center*) 76

A 1920s LNER poster advertising train travel as a means of getting to the eclipse site. (*Reproduced with kind permission from the Science and Society Picture Library*) 80

Engraving of the emu below the night sky's emu. (*Photograph reproduced with kind permission from Ray and Barnaby Norris, www.emuinthesky.com*) 84

Various southern hemisphere constellations. (*Bevis atlas, plate 49, MAS*) 92

A globe, made by Greaves and Thomas, where the traditional constellations have been replaced by characters from Lewis Carroll's *Alice in Wonderland*. (*Photograph reproduced with kind permission from Greaves and Thomas, www.globemakers.com*) 97

The Brownies' stargazing badge. (*Reproduced with kind permission from Girlguiding UK*) 102

The constellation Hydra surrounded by various smaller constellations, from a set of nineteenth-century educational cards called 'Urania's mirror'. (© *National Maritime Museum, Greenwich, London*) 108

Joseph Wright's 1768 painting, *An Experiment of a Bird in an Air Pump*. (*London, National Gallery*) 111

The star Eta Carina close up. (*Photograph from NASA's Hubble Space Telescope*) 116

The southern sky as it was known to the ancient Greeks. (*Bevis atlas, plate 51, MAS*) 119

Illustration of the Milky Way. (*Image produced by Richard Powell; it can be found on the Atlas of the Universe website, www.atlasoftheuniverse.com*) 122

The constellation Ara. (*Bevis atlas, plate 46, MAS*) 124

The Dunhuang manuscript showing the northernmost constellations around the pole star. (*Reproduced with kind permission of the British Library*) 129

The constellation Auriga. Illustration. (*Bevis atlas, plate 12, MAS*) 132

The constellation Ophiuchus and Serpens. Illustration. (*Bevis atlas, plate 14, MAS*) 134

The constellation Unicornis. Illustration. (*Bevis atlas, plate 39, MAS*) 135

Illustration from Edwin Dunkin's popular science book of 1869, *The Midnight Sky*, showing the Milky Way over London. (© *National Maritime Museum, Greenwich, London*) 138

The constellation Orion. Illustration. (*Bevis atlas, plate 35, MAS*) 146

The constellation Cetus. Illustration. (*Bevis atlas, plate 34, MAS*) 150

The Hertzsprung–Russell diagram, showing the position of a few well-known stars. (*Diagram courtesy of NASA*) 157

Orion nebula M42. (*Image from NASA, the Hubble Space Telescope and C.R. O'Dell, Vanderbilt University*) 162

The constellation Leo. Illustration. (*Bevis atlas, plate 26, MAS*) 165

This depiction of the 1833 meteor storm comes from Adventist Joseph Harvey Waggoner's *Bible Readings for the Home Circle*, 1888. (*Reproduced here with kind permission from the Armagh Observatory*) 167

The orbital path of a typical comet. Illustration. (© *Greg Smye-Rumsby*) 172

'The Female Philosopher Smelling out the Comet'. Illustration. (*Reproduced with kind permission from the Ohio State Cartoon Research Library*) 179

The constellation Cassiopeia. Illustration. (*Bevis atlas, plate 10, MAS*) 186

The constellation Andromeda. Illustration. (*Bevis atlas, plate 20, MAS*) 194

The Andromeda galaxy. (*Photograph reproduced here with kind permission of Robert Gendler*) 196

The constellation Perseus. Illustration. (*Bevis atlas, plate 11, MAS*) 198

The constellation Corona Australis. Illustration. (*Bevis atlas, plate 47, MAS*) 206

Diagram showing the tilt of Earth and the celestial equator. (© *Greg Smye-Rumsby*) 209

The constellation Eridanus. Illustration. (*Bevis atlas, plate 36, MAS*) 211

The constellation Argo Navis. Illustration. (*Bevis atlas, plate 40, MAS*) 223

The constellation Gemini. Illustration. (*Bevis atlas, plate 24, MAS*) 233

The Porter Garden Telescope. (*Photograph reproduced with kind permission from Telescopes of Vermont, www.gardentelescopes.com*) 239

The sky north of the ecliptic. Illustration. (*Bevis atlas, plate 50, MAS*) 246

John Flamsteed's comic horoscope of the Royal Observatory. (© *National Maritime Museum, Greenwich, London*) 253

The constellation Taurus. Illustration. (*Bevis atlas, plate 23, MAS*) 255

The Pleiades open star cluster. (*Photograph reproduced here with kind permission of Robert Gendler*) 258

The constellation Virgo. Illustration. (*Bevis atlas, plate 27, MAS*) 262

Star charts. Illustrations. (© *Greg Smythe-Rumsby*) Plate section, 1–8

Index

Notes: Where more than one page number is listed against a heading, page numbers in **bold** indicate significant treatment of a subject. Page numbers in *italic* indicate illustrations are present